The Anatomy of a Business Strategy

The Johns Hopkins / AT&T Series in Telephone History

Louis Galambos
The Johns Hopkins University
Series Editor

The integrated Bell telephone set in 1882. This unit was standardized by the Charles Williams, Jr., electrical manufacturing shop in Boston. It was produced by the Western Electric Company of Chicago after its acquisition by the American Bell Telephone Company. In one housing the set combines a magneto call bell, a Bell receiver on an automatic switch hook, a Blake transmitter, and a box in which a local power battery was stored.

The Anatomy of a Business Strategy:
Bell, Western Electric, and the Origins of the American Telephone Industry

George David Smith

The Johns Hopkins University Press
Baltimore and London

© 1985 The Johns Hopkins University Press
All rights reserved
Printed in the United States of America

The Johns Hopkins University Press, 701 West 40th Street,
Baltimore, Maryland 21211
The Johns Hopkins Press Ltd, London

The paper in this book is acid-free and meets the guidelines for permanence and durability of the Committee on Production Guidelines for Book Longevity of the Council on Library Resources.

Library of Congress Cataloging in Publication Data

Smith, George David.
    The anatomy of a business strategy.

    (The Johns Hopkins / AT&T series in telephone history)
    Includes index.
    1. American Telephone and Telegraph Company—History. 2. Western Electric Company—History. 3. Telephone—United States—History. I. Title. II. Series.
HE8846.A55S65   1985   384.6'065'73   84-23419
ISBN 0-8018-2710-8 (alk. paper)

For my mother

Contents ❧

Editor's Introduction ❧

In the past century Americans have built the foundations for a new society. Reflect for a moment on just a few of the major changes that have taken place. A society with virtually no administrative state has created one of the largest, most complex sets of governmental institutions in the world. A nation that had historically avoided involvement in the foreign affairs of Europe's leading powers has become the center of an elaborate alliance system that reaches into every corner of the world. An agrarian society has become over-whelmingly urban and industrial. An economy characterized by relatively small, local, or regional businesses has been transformed by the rise of great modern corporations with operations that span the entire nation and embrace markets and sources of supply around the world.

As a result of these changes, we have gradually become accustomed to very large, complex administrative systems. Despite periodic outcrys and symbolic political assaults against public bu-reaucracy, most Americans have learned to work in administrative settings and to think instinctively in terms of a collectivity or group. Planning, for instance, is a normal function of modern organized life. We simply assume that every public and private group will plan ahead, looking out for its own future in a systematic fashion. System itself is an accepted part of our lives. Orderly and rationally con-trolled patterns of behavior are so commonplace that few of us can conceive of doing things in a different way. We expect to have our performances measured. Numbers are intrinsic to our techniques for understanding society and evaluating everything from aspirin to our nation's defense system and our children's work in school.

So deeply embedded are these ways of thinking that most of us have difficulty imagining what the attitudes and daily lives of our

great-grandfathers were actually like. Then, large-scale administrative systems were a rarity in America. In nineteenth-century
society, small and relatively informal groups were the most common
forms of social organization. Methods of control were individual, and
formal planning was an exceptional and far less complex exercise
than it is today. Personality loomed large in the country's public and
private life. In politics the successful man wove elaborate networks of
personal party ties, which he used to climb to the top of state or
national affairs. In business a man's personal reputation, and hence
his credit rating, was often his most important single asset. It was
possible to go into business without having much capital, but you
could not make it without credit.

The average business of the post–Civil War era was small and
was directed by an entrepreneur who dealt on a face-to-face basis
with most of those who were essential to his firm's progress. If he
needed credit, he sought it from persons who knew him, were
familiar with his business, and had some solid grounds for believing
he would be able to pay his bills. When he sold goods or services to
customers or to wholesalers, he, or perhaps his partner, knew the
individuals with whom he was dealing. While the normal owner-
entrepreneur of the 1870s and 1880s could hardly have kept track of
all of his workers, he had to know and trust his foremen and
supervisors. Frequently he left the job of choosing the work force,
setting wages, and determining the conditions of work in the hands
of the foreman, who subcontracted to produce the company's products.

This type of entrepreneurial firm was especially well suited to
doing business in a rapidly changing economy of the sort that existed
in the United States in the late nineteenth century. It was a very
flexible, adaptable style of organization, attuned to innovation in
the race to exploit the nation's abundant natural resources. There
were few impediments to change within this type of business; the
route from business decision to business action was short, as was the
planning or expectational horizon of the firm. Both were bounded by
the personality of the entrepreneur. Although the failure rate was
high—especially during the major depressions and financial panics
that periodically interrupted the country's development—there were
always new entrepreneurs eager to take the place of those who had
faltered.

Out of this beehive of capitalistic activity there emerged, in the last two decades of the century, new forms of business organization that would come to characterize our twentieth-century economy. The large corporation gradually supplanted the entrepreneurial firm in manufacturing, in distribution, and in communications, as it already had by 1880 in the nation's railroad industry. The traditional style of small business did not wither away; it is still an important factor in our economy today, particularly in new industries, where it provides an innovative impulse much as it did in the nineteenth-century economy. But most of our goods and services now come from very large corporate businesses. Using modern technology and producing for mass markets, these corporate giants dominate our economic environment by dint of their efficiency in large-scale production and distribution. In the latter part of the nineteenth century, however, they were on a much less secure footing. Then, they were in flux, finding it necessary, for instance, to develop new kinds of administrative systems that would enable them to control effectively their far-flung and varied operations. In this setting personality gradually gave way to business administration. A single person could no longer deal directly with all of the firm's customers, sources of credit, or supervisory personnel. With large size came specialization of function. Slowly and hesitantly big business experimented with new techniques, searching for appropriate means of communication, control, and planning.

In retrospect, the process of administrative consolidation appears to have been ad hoc and haphazard. Indeed it was. The businessmen involved were exploring new territory. There were hardly any models for them to imitate, no organizational specialists to consult, no theoretical or practical literature charting the path to efficient administration. Frequently they were constrained by shortages of capital. Their administrative resources were unduly strained by the tasks of keeping the corporation profitable on a daily basis while they developed viable systems to sustain its operations over the long run. They were pioneering in business administration in a fluid technological and economic setting that frequently threatened to undermine their most innovative efforts.

One of the significant strategies for long-run growth adopted during these years was that of vertical integration—that is, the combination in one firm of sequential steps in the business process, from the acquisition of the raw materials, through the distribution of its goods or services. Firms manufacturing for the national market acquired their own distribution facilities and sometimes their sources of raw materials. In some cases distributors integrated backward into manufacturing. The corporation thus internalized and—if it succeeded in administrative consolidation—brought under common ownership functions that had previously been beyond the firm's control. In the same manner, large corporations eventually internalized the process of technological development by establishing industrial laboratories and engineering departments. Once internalized, these activities could be the subject of planning and of systematic techniques of administrative control. But first of course the corporation's leaders had to learn how to organize and systematize their operations.

It is in the midst of this significant transition in business organizations and techniques that George David Smith's study *The Anatomy of a Business Strategy: Bell, Western Electric, and the Origins of the American Telephone Industry* places the reader. As the author demonstrates, the process of vertical integration had certain unique characteristics in the case of the early Bell enterprise. This particular business strategy was influenced decisively by the company's competitive environment, its meager capital resources, and its need to acquire administrative talent. Only gradually did the company overcome these initial problems.

Smith's book—a volume in the Johns Hopkins/AT&T Series in Telephone History—provides us with a superb analysis of the manner in which the firm became an integrated corporation and devised appropriate modes of administrative control. One element in this transformation was the development of standardized contracts. Another was the coming to power of the full-time professional manager. In the case of the Bell Company, a small group of part-time businessmen had successfully launched an undertaking that proved to be such a success that it, in effect, outgrew their talents. They were replaced by professionals who reshaped the business and provided it with the rudimentary tools of modern management. Boundaries between agents of the firm were carefully and explicitly defined. Technological innovation was vigorously promoted within the enterprise and technical standardization encouraged throughout the emerging national network. Within the Bell Company, responsibilities of the employees were explicitly delineated, and potential conflicts of interest defined and, so far as possible, eliminated. Costing and pricing policies were systematized. Theodore N. Vail and subsequent generations of Bell managers solved these problems and eventually created a new form of integrated and consolidated business enterprise. Functions such as technical innovation were successfully internalized, as was manufacturing. If this process took several decades to complete—as Smith shows it did—it was to provide an enduring base for what would become the world's largest business. The Bell System would be one of those great modern organizations that would shape American attitudes toward planning and large-scale administrative systems, and even today, following the breakup of the national network, the American Telephone and Telegraph Company is a vertically integrated corporation.

Louis Galambos

The AT&T Corporate Archives ❧

It is the policy of AT&T to preserve all company records and artifacts that document its history—the evolution of its structure and organization, the development of its products and services, and the evolution of corporate policies. Therefore, in addition to record retention requirements imposed by law and retention standards adopted by individual departments for their own purposes, the potential corporate historical value of records will be taken into account in the retention and disposal process. The company's central archival organization has responsibility for assessment of the historical importance of all company records and artifacts and the designation, accession, and preservation of archival records.

The company policy is to stimulate scholarly awareness and use of the materials. AT&T has undertaken to make these records and artifacts available for corporate and approved scholarly reference and use through a systematic program of accessioning and cataloguing, by the establishment of adequate reading room facilities, and by direct preparation of publications.

Preface and Acknowledgments ❦

This book considers a single business decision that had significant ramifications for America's economic and business history. I attempt to explain how the American Bell Telephone Company's decision to acquire the Western Electric Manufacturing Company grew out of a complex array of evolving circumstances and what difference that decision made in the fullness of time. As the organizational keystone of a pioneering effort to integrate large-scale capitalist enterprise, the acquisition of Western Electric was an important business event of its day. The durable success of the relationship between Western Electric and the dominant firm in U.S. telecommunications—a relationship that has survived considerable political, legal, and economic controversy—merits a close study of the origins and rationale of the decision that forged the relationship.

Detailed research in primary sources and the writing of this book have required the full cooperation, as well as the financial support, of the American Telephone and Telegraph Company. In the latter part of 1980 AT&T asked me to undertake the history of the acquisition of Western Electric. I had already done some work as a consultant to the company, and by early 1982 I was able to write a preliminary draft of this study. Then the turmoil surrounding the company's reorganization in the wake of the great antitrust settlement of that year left the work in abeyance. When in 1984 I was asked to revise and refine the manuscript for publication, the early history of the company had taken on a surprising relevance, even though current events seemed to be propelling AT&T away from much of that past into a radically new future. The relevance, to my mind, lies in the news that this book bears for the company and the country about an entrepreneurial part of Bell's past that has been largely forgotten.

Like all great business enterprises, AT&T has a historical memory that conditions its sense of corporate purpose, strategic outlook, policies, and protocols. At the center of AT&T's corporate tradition is its founding legend, the heroic tale of Alexander Graham Bell's stubborn pursuit of a practical means to transmit the human voice over distance by means of electricity. The story of the invention of a practical telephone, which became the basis of the modern telecommunications industry, is but the first of many stories of scientific and engineering achievement in the Bell Telephone System, stories that generally exalt the development of technology as a means to corporate growth and, not incidentally, social improvement.

Perhaps no less important to AT&T's self-image has been a body of tradition stemming from the company's development into a huge managerial and engineering system supplying and operating the complex technology for most of the telecommunications in the United States. This is the "universal service" corporation that emerged under the leadership of the great industrial statesman Theodore Vail between 1907 and 1919. It was during that time that AT&T consolidated its technological, financial, and managerial control over American telephony while reaching accommodation with government regulation. One important result of this process was the development of a powerful corporate ethos defined by the systematic improvement and expansion of a national telephone service administered through tightly integrated bureaucratic and technical organizations, all supported by fairly predictable, regulated rates of return.

The tradition of AT&T as "invented" by Theodore Vail has been publicly articulated in a recent, thoughtful memoir by former company vice president and assistant to the chairman Alvin von Auw, whose *Heritage and Destiny* (1983) celebrates the spirit of the Bell Telephone System in its passing. Von Auw's reading of the corporate history vividly reflects the attitudes emphasizing stability, order, security, and predictability that had become very deeply embedded in the Bell System's corporate culture since Vail's regime. In recent years, a series of external political events—culminating in the company's decision to divest much of its telephone business in the antitrust consent decree of 1982—have done great violence to the corporate values based on the history of a regulated monopoly. But now that the combined pressures of technological and political forces have decisively transformed the market structure of the industry (creating a less certain, more competitive world for AT&T and Western Electric managers), it seems wise to pause and look back at the half-forgotten elements of a more distant company past.

This book provides a small window into a time when a much younger Theodore Vail joined the infant Bell Telephone Company, a very high-risk enterprise founded on an uncertain technological base in an ill-defined and unregulated market. It is a history of entrepreneurial behavior, competitive strategies, loosely bound technologies, and fast-changing organizational structures. It is a history of creative response to rapidly evolving opportunities and threats. It is a history pertinent to an era of corporate renewal in an increasingly unpredictable business environment.

It is above all a history; and whatever relevance to contemporary events it may have, it remains the historian's duty to treat the past on its own terms. Otherwise, the history becomes less a useful perspective on the past than an anachronistic reflection of the transient concerns of the moment. I have tried to recapture the perspective of the decision makers as they looked forward to uncertain outcomes. To provide some assessment of the long-term results, I have carried the discussion of some of the consequences of the acquisition to 1915, by which time the main outlines of the modern relationship between Western Electric and the Bell Telephone System were clear.

The primary audiences for this work are students of professional management, business and economic historians, and the company's own managers, in other words, audiences that, from the company's point of view, are both sophisticated and significant. Addressing these audiences required an approach to the sponsorship of corporate history rare among American corporations. AT&T provided me with complete and open access to its rich historical collection of primary sources and encouraged me at every stage to develop the story entirely in accordance with the canons of historical scholarship. AT&T granted me unqualified editorial and interpretive freedom throughout and bears no responsibility for any opinions or conclusions expressed in the text.

I am particularly thankful to AT&T personnel who gave me intellectual and moral support in the drafting of this book. Philip Haff promoted the work and secured its funding with good-humored determination. Robert G. Lewis provided critical advice on the text and was largely responsible for bringing it to publication. Robert Garnet, who worked on his own history of AT&T from a different perspective, was a valuable and amiable colleague. Alan Gardner, a consulting researcher for AT&T, was a stringent critic as well as a strong resource on matters of fact and interpretation.

Two eminent historians have been important in the shaping of my work. Louis Galambos, professor of history at The Johns Hopkins University, provided me with incisive critical reviews of successive drafts of the manuscript. His well-focused editorial advice and questions helped me render the thematic content of the book clearer than it would otherwise have been. James P. Baughman, now of the General Electric Company, was a primary consumer of my research as he prepared for expert testimony on the history of the Bell System. He was a strong advocate of conducting serious scholarly work on AT&T's history and of my work in particular. I am ever grateful to him for the lessons he taught me about business.

Others who contributed useful criticism and advice include: Laurence Steadman, of Steadman and Coles in Boston; Fred Cardin, David Kiser, and Robert Shay, former colleagues at the Cambridge Research Institute in Cambridge, Massachusetts; Neil Wasserman, of the Harvard Business School; Thomas Huertas, of Citicorp in New York City; Margaret B. W. Graham, of Boston University; Bernard Bailyn, of Harvard University; and Peter Temin and Mel Horwitch, of the Massachusetts Institute of Technology.

Thanks also go to Millie Ettlinger, Richard Priest, Deborah DeFago, and Claudia Penna, of AT&T; Young-Hi Quick and George Gray, of Western Electric; George Schindler and his staff at Bell Laboratories; and Eric Hanin, Robert Heffley, Jane Radcliffe, William Quinn, Judy Uhl, Cynthia Rose, and Joan Terrell, all former colleagues at the Cambridge Research Institute. Special thanks to my partners at the Winthrop Group, Inc., who have provided commentary and good cheer along the way.

The Anatomy of a Business Strategy

CHAPTER 1 🐦

Overview and Context

THIS IS A HISTORY OF A BUSINESS DECISION made more than a century ago. The decision of the American Bell Telephone Company to acquire a majority interest in the Western Electric Manufacturing Company and its strategic ramifications have had profound and lasting impact on the telecommunications industry. Western Electric's role in the modern Bell Telephone System has received a great deal of attention from those scholars, attorneys, and government officials interested in the economic and legal consequences of vertical integration; and yet there has been little serious inquiry into the historical origins of the arrangement.[1] This is a significant oversight, because although the Bell Telephone System became historically identified with the development, provision, and maintenance of telephone service, the distinctive feature of the structural evolution of the Bell enterprise was its early, primary emphasis on the production and distribution of hardware. Thus, to understand the organizational basis for what was to become the world's largest business and technological enterprise, it is necessary to understand the circumstances and process that led to the consolidation of interests between Bell and its historic manufacturing arm, Western Electric.

The early history of the Bell enterprise is important because the company's initial patterns of development have continued to have a decisive influence upon the business's tightly integrated vertical structure[2] and its decentralized operating structure even to the present day. The Bell System's early history, moreover, stands as a significant case in the emerging pattern of late-nineteenth-century big business consolidation and the rise of the modern corporation. Since the history of the Bell System is typical of the more general pattern of big business development in many respects but unique in others, the precise contours of its development are matters of abiding concern to industry economists, policy makers, and management. It is hoped that this small book will be a useful contribution to that concern for understanding.

Alfred D. Chandler, Jr., in his grand synthesis of the development of American business structure,[3] shows how strategies of vertical integration (along with the combination of competing firms) began to emerge on a large scale in the 1880s. Both the changing nature of the market and technological innovation influenced this significant process of organizational change. Citing in particular the strategies of manufacturers who employed continuous-process machinery or who required highly specialized modes of marketing and distribution, Chandler finds that vertical integration was a rational managerial and economic response to technological change and market opportunity. Vertical integration helped unleash the technological potential of mass production for a mass market—a potential made possible by increasing urbanization and by the nation's maturing railroad transportation and telegraph communications systems.

The firms that generally pioneered in strategies of integration forward into distribution were capital-intensive makers of products that fall into four general categories: (1) mass-produced, low-priced, packaged semiperishables, which were manufactured by large-batch or continuous-process technology; (2) processed perishables for distribution in national markets; (3) mass-produced machines to be sold in mass consumer markets, which required specialized marketing services; and (4) high-volume producer goods, standardized but technologically complex enough to require demonstration and service.[4] Not all producers in these categories chose to integrate, but those that did not were eventually eclipsed by the more innovative firms which reaped the competitive cost advantages of large-scale, integrated enterprise.

As a look at these four categories makes clear, the basic pattern of forward integration during the 1880s involved producers seeking to combine mass distribution with their capacity for mass production. Producers of semiperishables (American Tobacco, Diamond Match, Quaker Oats, Campbell Soup, Procter and Gamble, Eastman Kodak)[5] created national and international networks of sales offices in order to brand, advertise, and manage flows of goods from wholesalers to retailers. Processors of perishable products (Swift, Armour, Pabst)[6] by-passed the wholesalers altogether, selling directly to retailers. Makers of mass-produced machines and of high-volume producer goods (Singer, Westinghouse, and other firms in the electrical industry, along with Babcock & Wilcox and McCormick)[7] found that they had to provide direct demonstration and maintenance services—and often credit—to their customers. The most successful of these firms created wide networks of sales offices, tightly coordinated and controlled from the center of the organization.[8]

Backward integration by distributing firms was far less common. It was only necessary when the firms had problems obtaining goods in the open market at the quantity, quality, and price desired. Once stable sources of supply were assured, most distributing firms were likely to sell out their production facilities or else maintain only a passive concern with them.[9]

Some large enterprises nonetheless found it imperative to integrate both forward and backward in order to support their large-scale operations in national markets. The best-known, indeed classic case is the Standard Oil Company's movement backward from refineries into shipping, forward into marketing, and then further backward into the extraction of crude oil from its source.[10] By the time John D. Rockefeller and his associates had, in 1882, used a trust to solidify the legal basis of their control over some forty refineries, they were already building long-distance pipelines.[11] This new transportation technology had the potential to slash the costs of shipping crude oil substantially below those incurred by using the rails. This technological innovation triggered a further consolidation of refineries through merger, which in turn required greater centralization of administration and a concomitant development of systematic and "professional" management. All this was necessary in order to realize economies of scale from the firm's greatly expanded volume. Once these managerial problems had been solved, Standard

Oil moved forward into wholesaling to ensure a more rapid flow of high-volume output into the market. The combine also moved backward into drilling, a defensive move taken to guarantee its sources of supply against a possible combination of producers.[12]

Vertical integration of the giant, national corporate enterprises that emerged after 1880 generally involved this kind of quest for scale and efficiency, influenced by the push of mass-production technology and the pull of widening markets. Successful growth through merger was sustained by the elaboration of formal managerial systems and in many cases by the vertical integration that enabled the firm to obtain supplies as efficiently as it now manufactured them.

Seen in this context, the early development of the Bell Telephone System was unique in many regards. Neither large-scale mass production nor mass distribution was involved in the very early years of the telephone industry. There was no giant horizontal combine, no trust of the Standard Oil variety. There were no maneuvers by well-capitalized opponents seeking to checkmate their foes by controlling vital goods or sources of raw materials. And there was little in the way of a conscious quest for economies of scale through the aggregation and centralization of production.

In other regards, however, Bell's evolution in the early years shared certain elements common to the big business enterprises of the late nineteenth and early twentieth centuries. From the beginning, the Bell entrepreneurs (much like their forerunners in the telegraph industry)[13] sought to control their sources of supply and channels of distribution for a national market. They did so in response to the strategic imperatives of patented technology and growth and to the underlying threat—even when their patent monopoly may have seemed secure—of competition. As with many new technologically based enterprises, patent protection and the effort to promote and manage technical innovations became significant factors very early in the history of the telephone. Gradually, too, expanding markets began to create problems as well as opportunities for the Bell System. By the time the decision to combine Bell and Western Electric was made, in fact, large-scale (if not mass) production was a serious consideration, as were the questions raised by distribution to rapidly growing urban markets throughout the

United States. These matters take us too far ahead of our story, however; now we must look back to the first years of the telephone and the origin of the Bell System.

· I ·

The Bell Company[14] began its commercial life in 1877 as the holder of potentially controlling patents over a potentially valuable, though as yet commercially undeveloped, technology. The founders of the business might have developed their patent in a variety of different ways, perhaps by manufacturing and marketing the telephone themselves. Indeed, one of their earliest formal corporate charters provided for the Bell Company "to manufacture and sell telephones and their appurtenances and to construct, maintain and operate lines . . . throughout the United States. . . ."[15] They might also have embarked on an ambitious program of organization to strive to realize Alexander Graham Bell's early conviction that central offices could be built to establish "direct communication between any two places in [a] city" and that wires between cities would make it possible for "a man in one part of the country [to] communicate by word of mouth with another in a different place."[16]

In 1877, however, all this was out of the question because the Bell Company neither had nor seemed likely to acquire the capital or the technology needed for such an extensive business. Instead, the patent holders cut their plans to fit their limited resources. They embarked on a policy of licensing local entrepreneurs on a geographically exclusive basis to provide simple, point-to-point service to end users. To these forerunners of the modern telephone operating companies telephones, manufactured under the Bell patents, were leased, not sold. To provide the licensees with equipment, the Bell Company turned to independent manufacturers of electrical equipment—at first just one and later several—which were licensed to produce the various pieces of telephone hardware on which patents were held. In the initial phases of its business, Bell earned no income on either the provision of telephone service or the manufacture of telephone equipment. Income went instead to the licensed operating agencies and producer(s), respectively. Bell made its profits from leasing telephones and from charging royalties on the production of patented hardware.

The salient managerial problem arising from this licensing policy was the monitoring and coordinating of the production and distribution of telephone equipment.[17] It was important for the patentees to meet their operating agencies' demands for equipment. They had to supply a growing market with reliable hardware at reasonable cost while at the same time reserving to themselves tight control over the patented technology. If Bell expanded output but somehow weakened its control of the technology, the firm would be seriously endangered. If, however, Bell did not provide the operating agencies with good-quality equipment in the amounts that they needed, the licensees would be encouraged to look elsewhere for these goods. Bell had to steer between these two rocks while carefully husbanding its limited capital. To those familiar with the present-day company, it may be difficult to conceive of how tenuous and circumscribed the firm's prospects were a century ago.

The Bell Company's first manufacturing arrangement reflected the firm's modest beginnings and uncertain prospects. It was based on a very informal and closely personal relationship between the holders of the patents (the acoustical expert Alexander Graham Bell, his research assistant Thomas Watson, and two fathers of deaf students whom Bell tutored) and Charles Williams, Jr., an electrical machinist whose shop had been the site for Bell's early telephone experiments. This informal, small-scale arrangement could not endure, however, in the face of rapidly growing demand and the advent of heavy competition from a well-financed, rival patent holder. In two years the Bell Company was forced to expand its capacity by licensing four additional, geographically dispersed electrical manufacturers to produce auxiliary apparatus on which it held patents. By 1881 a new approach to production was conceived: manufacturing was to be consolidated in one firm with several branches, in which the Bell Company would take a majority equity position for purposes of control. The target for acquisition was the Western Electric Manufacturing Company, the largest electrical equipment manufacturer in the country.

Obviously, much had to have changed since the Bell Company's humble beginnings for such an acquisition to take place. In the first place, Bell's formidable rival, Western Union, had with-drawn from competition in late 1879 following one of the more remarkable settlements of a patent suit in the history of business. By that time the Bell Company had grown from a small, unin-corporated, thinly capitalized group of patent holders into a more professionally managed company of substantial size able to coordi-nate a more extensive, much broader, more complex range of operations on a national front. Then, with its principal competitive threat under control, the value of the Bell Company's patents increased greatly, allowing the firm, for the first time in its existence, to raise enough capital to support an ambitious, long-term develop-ment of its business. From that point, the expectational horizons of the Bell interests lengthened and widened considerably. By 1880 the company determined to pursue, on a grand scale, the technological and business development of the central exchange and long-distance transmission—all of which had been but a gleam in Alexander Graham Bell's eye just two years before.

Accordingly, a new American Bell Telephone Company embarked on a series of structural, legal, and financial plans that by 1885 gave the company an ownership position in most of the firms connected with the industry. Bell moved forward into the dis-tribution of telephone service by consolidating, and then taking equity in, its operating licensees. It created a subsidiary to build long-distance lines after attempts to license others to do so. But the first major development in this program was the Bell management's decision to acquire and consolidate its own manufacturing capacity. It was in this period, with the integration of firms responsible for a wide range of functions along the vertical axis—from the develop-ment and production of telephone equipment to its installation and operations on a national scale—that the Bell Telephone System was born.

· II ·

The transformation of the Bell enterprise from a simple patent franchiser (which arranged for the production and distribution of telephone equipment) into a vertically integrated corporation of national scope is the essence of our story. The following chapters examine the precise contours of this strategic and organizational history with the decision to acquire Western Electric as the focus. This focus is crucial because the business began as a production problem, and its every subsequent development hinged on being able to secure a controllable source of adequate supplies. But before we can comprehend fully the series of decisions that preceded the acquisition of Western Electric, it is necessary to delineate the general and specific contexts of the Bell entrepreneurs' decision-making process. First, let us look at the broad business climate of the period.

Despite early conceptions of the telephone as a mere novelty or curiosity, its primary market was excellent. The telephone had great potential value for business establishments and wealthy house-holders in urban centers, if only as an efficient substitute for the short-distance, point-to-point telegraph. The United States was still a predominantly rural and agricultural nation, but the country's urban population was growing rapidly, from 9.9 million in 1870 to 14.1 million in 1880. The number of business concerns was also increasing at a fast pace, from 427,000 to 747,000 during the same period. The outlook for the telephone was further enhanced by a rising general level of prosperity as gross national product per capita swelled from $170 to $205 during the decade.[18] The market, in other words, was there; the question was whether the Bell interests could mobilize the capital and the organizational wherewithal to exploit it.

The decade of the 1870s was an expansive but erratic period in American industrial development, characterized by two phenomena particularly obvious to contemporary businessmen. The first was the continuing and long-term deflation of prices since the panic and depression of 1873; the second was the wide fluctuation in the level of business activity. That the economy was embarked on a long-term cycle of unprecedented high output, technological progress, and general prosperity was not sufficiently clear to allay the fears of businessmen about the uncertain prospects of their particular firms in a volatile short-term economy. Defensive strategies among manufacturing concerns, manifested in attempts to control production, prices, and distribution through business associations, or cartels, were common. While in the abstract, competition was regarded as the cornerstone of free enterprise, price competition was anathema to those caught up in it.[19] The Bell Company's situation was no different in this regard. In its dealings with its operating licensees, Bell countenanced no competition at all, granting to each an exclusive privilege to distribute its telephones in particular markets. In its dealings with apparatus manufacturers, Bell took a more mixed (or ambivalent) approach, encouraging the manufacturers to compete for the business of the operating agencies while forbidding them to vary their prices from one to the next. Production, while rationally bound only by the rate of growth of Bell-licensed agencies, was further checked by the capacity of licensed manufacturers and their willingness to increase output in an uncertain world.

The 1870s also saw the first important extension of the electrical manufacturing business beyond its early applications in telegraphy and electroplating. As the applications increased, the firms that emerged to exploit them were organized around patent claims. Typical of the organization of all the major firms in the electrical industries, telegraph and telephone company organization "crystallized around patent rights," and so whoever desired to enter or sustain business in either field had to come to terms with the holders of significant patents.[20] The early commercial development of telephony proceeded amidst rivalry between two corporate entities, each claiming to hold the controlling patents on telephone technology. When Western Union and the National Bell companies came to an agreement in November 1879, the issue was not entirely laid to rest. The Bell interests looked forward to challenges not only to the telephone instrument but also to the important apparatus growing up around the rapidly developing exchange technology. Patents were the lifeblood of the business. Survival (in this as well as in most emerging high-technology businesses of the era) required almost obsessive attention to patent claims wherever they arose and to continuing development of one's own patent base.

Set against this general business background were the basic business and technological concerns of Bell management—concerns that made up the immediate ideational framework around which they organized production and distribution. Emerging from the entrepreneurs' position as holders of potentially valuable patents was a constant set of objectives to underpin the otherwise turbulent succession of organizational arrangements that marked the early years of the business. Four fixed objectives which became the cornerstones of company policy can be defined in hierarchical importance as follows. First was the entrepreneurs' overriding desire to expand their markets under the strongest possible protection of patents; second was their corresponding need to achieve a relatively high volume of business and keep it growing; third was their belief that to maintain a strong position in the industry's technology, they had to innovate; and finally, stemming from their other objectives, was their insistence on high standards of quality in production and service. These four fixed objectives run throughout our story like interwoven threads of policy, binding together the evolving elements of company strategy and structure.

Although some things remained fixed, others changed. In the first four years of its life the Bell Company's strategic and structural evolution was affected by challenges and constraints that shifted over time. It is crucial to identify these challenges and constraints clearly because they placed increasing pressure on the company to take greater control over the technical development and production of telephone equipment. The critical challenges to the business were threefold: (1) an ever-increasing rate in the growth of demand; (2) competition for control of the market from Western Union (from late 1877 to late 1879); and (3) ongoing developments in the technology from actual or potential competitors throughout the period. The constraints limiting the range of responses to these challenges were (1) a chronic shortage of capital; (2) a paucity of administrative resources; (3) limited technological resources; and (4) limited productive capacity. These challenges and constraints, which were two sides of the same coin, varied in importance and intensity over time, but their presence had formative impact on the company's long-term institutional development.

Just how these challenges and constraints altered over time is an important part of our story. Their ebbs and flows, and their mix, were the daily concerns of the Bell pioneers, whose assessment of just where they stood with respect to these factors guided their organizational decisions. Their decision to move from their unitary reliance on Charles Williams, Jr., to a multiple-licensing arrangement for their manufacturing requirements and their decision later to reconsolidate and acquire a manufacturing capability were made under very different sets of challenges and constraints. We shall examine in detail the historic process that made this difference. At the same time, we shall discover that the same process did not alter the fundamental concerns of the Bell Company for patent protection, increasing volume, technological supremacy, and high production standards. These were the anchors for decisions wrought in otherwise shifting circumstances.

· III ·

The early history of the Bell Telephone System is one of entrepreneurial success, and so it might be tempting, in retrospect, to regard the company's formative financial, organizational, and managerial decisions as all too rational, all too obvious. But in the eyes of the early Bell entrepreneurs this was certainly not the case. They frequently disagreed as to what should be done, and they saw many alternatives to the specific measures adopted. Each major decision seemed full of peril for a relatively small firm just getting its feet planted. The reader should bear this in mind, even as the historical narrative imposes its own highly rational order upon a very human and often confusing process.

The solutions adopted by the early Bell entrepreneurs, moreover, were to a large degree inventions. During these years there were no generally accepted models of large-scale organization outside the relatively new railroad and telegraph industries; in most industries the processes of horizontal and vertical integration were just getting under way. The corporate form of organization, still an uncommon device for raising and allocating capital, was not yet highly developed as a business form.[21]

Nor was there a bureaucratic caste of salaried managers, in the modern sense, whose task it was to plan and systematically administer large-scale, integrated enterprises. There were in the United States no business schools imparting formal organizational wisdom and no general guides to "best practice" in business development. There was no engineering profession prescribing systematic relationships between forms of technology, modes of production, and institutional arrangements.[22] The rise of the Bell System was contemporaneous with the origins of other pioneering, technology-driven, big business corporations and the professions that they spawned. Like the others, the Bell enterprise proceeded in an ad hoc and experimental manner until a body of experience provided some basis for more systematic strategies, more elaborate structures.

The strategic decision to acquire Western Electric (notwithstanding all that is implied by the subsequent history of its consequences) was made without any idea of what would ultimately become the modern Bell Telephone System. The decision was not the product of the thinking of systems engineers or organizational economists: it was rather the culmination of five years' experimentation by creative and opportunistic nineteenth-century businessmen bringing products manufactured under their patents into an unregulated national market.[23] As the early Bell entrepreneurs moved (albeit purposefully and with considerable long-range vision) from one manufacturing arrangement to another, transforming their technology, management practices, and organizational structure along the way, they did so as pure empiricists. They did not grasp much of the potential for scale economies or transaction cost savings through the consolidation and centralized coordination of production, purchasing, and distribution. Nor did they apprehend the efficiencies that would later derive from coordinated innovation and standardized implementation of technology. They saw their business far less in terms of intricate, interactive, interdependent systems of technology and organization—that is a modern formulation of the nature of the enterprise—than in terms of discrete sets of hardware and companies, over which they strove to achieve strategic control.

Their problems were not simple, however, nor was their thinking naive. They faced technological and organizational problems of considerable scope and complexity for the times. Fortunately, the accumulated experience of other contemporary enterprises provided some guidance, some framework of understanding for planning the business. They and their licensees had before them, in the capital-intensive railroad and telegraph industries, vivid examples of far-flung, large-scale organization; technological complexity; and dynamic processes of traffic coordination, routing, and pricing that were immediately relevant to telephony. Among the Bell Company's early cadre of owners and managers were men with experience in the rails and the telegraph, and this was also true of the operating agencies. Gardiner Hubbard had some working knowledge of the shoe machinery business, which pioneered in the leasing of patented equipment, and Oscar Madden (who became the assistant general manager) came from the sewing machine industry, which pioneered in the coordination of flow of goods and services, first through patent-controlled franchises and later through vertically owned distributors.

In addition to their working knowledge of relevant business practices of other industries, the Bell owners and managers had before them the pressing example of Western Union's relationship with its various agencies and outlets, which ran the gamut from complete independence to complete ownership control. As for Western Union's relationship with Western Electric, it was clear that the two firms were not highly integrated. Nonetheless, it was also obvious that Western Union, in addition to reaping its share of Western Electric's dividends, enjoyed access to the manufacturer's numerous patents, technological expertise, and large (and responsive) productive capacity, all during its challenge to Bell's nascent telephone business. This lesson was not lost on the Bell interests after the patent rivalry was laid to rest. In the end, of course, it was to Western Electric that Bell turned when it began seriously to consider integrating backward into production.

This leads us to an important final consideration before turning to the details. It was Bell's good fortune, not careful planning, that made the prime electrical manufacturer in the country available for acquisition. Indeed, happenstance flows through much of the story that follows. Like the entrepreneurs who brought the telephone to market and their successors who turned it into an essential and ubiquitous commercial service, the Bell owners and managers often found their best plans undermined by unforeseen difficulties, their expectations overturned by unforeseen events, their hopes actually exceeded by unforeseen opportunities. It was also Bell's good fortune to have imaginative managers at the helm. Their ability to cope with the unexpected—to solve problems as they arose and seize opportunities as they appeared—makes the early history of the Bell System particularly instructive. The important thing for the reader to bear in mind is that the company's success was not predetermined. Its decisions were made in the face of great uncertainty and without recourse to modern conceptions of systems or microeconomic theories. Economic and technological forces helped to shape those decisions, but critical choices had to be made by individuals.[24] Every day that the company was in business, it had ample opportunity to fail, as well as to succeed.

Fortunately, the early Bell owners and managers left a rich documentary record. Since they did not rely on the telephone for discussions over policy or for exchanges of sensitive information, many elements of the decision-making process were vividly captured on paper. Ideas, even half-formed or untried, can often be described today with remarkable fullness. It is with adherence to this documentary record and with sensitivity to the nineteenth-century context in which the business developed that the following pages attempt to narrate and analyze in close detail the events associated with a seminal decision in the life of the firm that became the nerve center of American telecommunications.

Alexander Graham Bell (1847–1922) in 1876,
the year he patented the telephone.

Gardiner G. Hubbard (1822–97), a
Boston attorney and Alexander
Graham Bell's father-in-law, trustee of
the first Bell Telephone enterprise and
president of the first incorporated Bell
Telephone Company.

Thomas Sanders (1839–1911), a Massachusetts leather merchant who, along with
Hubbard and Bell, formed the Bell Patent
Association in 1875. He was treasurer of
the Bell Company until 1879.

Charles Williams, Jr. (1830–1908), whose
electrical shop on Court Street in Boston
developed the first viable commercial telephone instrument.

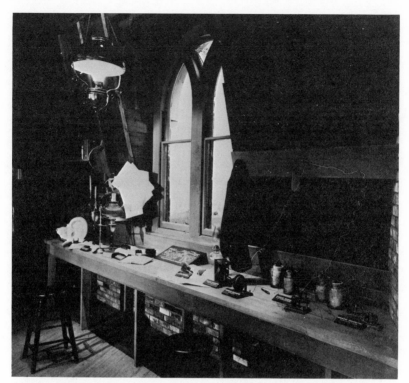

Alexander Graham Bell's original "laboratory" at 109 Court Street in Boston as re-created by the New England Bell Telephone Company.

The first commercial "box" telephone was developed in late 1876 and placed into service in 1877. The round, cameralike opening served as both receiver and transmitter, requiring awkward mouth-to-ear shifts.

17

The simplicity of the first commercial telephone instrument is evident from this photograph. The receiving/transmitting receptacle is mounted on a wooden block, behind which is attached an iron diaphragm. In transmitting, the speaking voice generated sound waves, which vibrated the diaphragm, setting up electromagnetic currents in a pair of copper induction coils attached to the ends of a permanent magnet. The currents were then transmitted through the wires (one a ground wire, the other a line wire). In receiving, the incoming electromagnetic impulses vibrated in the diaphragm, which emitted sound waves approximating the human voice.

Within a short time, the box magneto telephone underwent several changes in exterior design. Note the alterations in size and shape of the "mouthpiece," which also served as receiver.

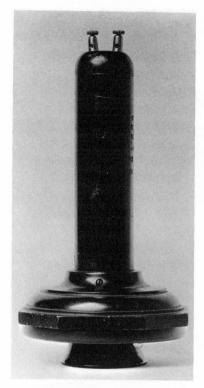

Almost immediately after the commercial introduction of the telephone, Charles Williams, Jr., began production of the hand telephone, which underwent many modifications in 1877 and 1878. The manipulative advantage of the "butterstamp" telephone, which served as both transmitter and receiver, made it more popular than the box, which was phased out of service by mid-1878.

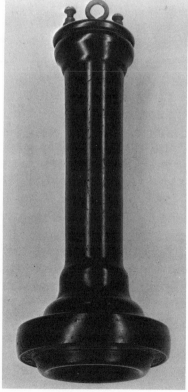

By the end of 1877, the hand telephone was encased in rubber, rather than wood, for better durability. The interior design of the hand-held model was essentially the same as that of the box phone, except that the magnet, a compound bar made of layers of magnetized steel, was straight and was attached to a single induction coil.

Watson's "Thumper" was the first Bell telephone signaling device, devised in June 1877. Here the box phone is equipped with a small hammer, which was activated by a spring-mounted exterior button. The striking of the hammer against the diaphragm sent an impulse through the transmission wire that was picked up as a "thumping" sound at the connecting telephone. Before the advent of the signaling device, the user tapped the diaphragm of the phone with a pencil.

Watson's "Thumper" was quickly superseded by the use of "call bells," electric bells that relied on batteries to power a transformer, which sent an alternative current over the subscriber's line. In mid-1878 Watson developed the "polarized call bell," shown here, the basic principle of which served the Bell telephone well into the twentieth century. The polarized bell consisted of two bells with a striker set between them. The striker was set on an armature polarized to oscillate when the alternating current passed through the winding of its electromagnet. It was like a small synchronous motor. The current was generated by a hand-operated crank located at each subscriber's station.

20

By early 1878 the hand telephone and crank-operated "magneto call bell" could be mounted on a single frame. The bell was activated by the crank set two-thirds of the way down the frame. At the lower end of the frame a hand-operated switch allowed the subscriber to change the telephone from the stand-by, or "calling," condition to the "talking" condition. The two telephones, each capable of transmitting and receiving, were installed so that the subscriber did not have to shift the instrument from mouth to ear. The installation of two phones at each subscriber station was Bell Company practice through the first half of 1878.

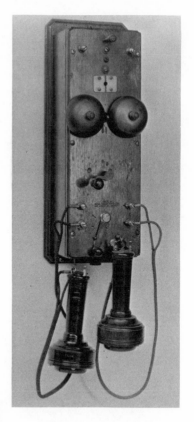

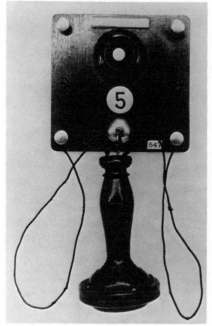

This "subscriber's station set" was used at New Haven and Meriden, Connecticut, sites of the first and second commercial switchboards, in early 1878. The circular device above the 5 is a push-button ringer.

The earliest telephone switchboards were fabricated, not by manufacturers, but by Bell operating licensees who experimented with ways to connect subscribers with more than one other station via a single subscriber's line. The first device resembling an exchange switchboard was adapted from a nighttime burglar alarm by E. T. Holmes in Boston in May 1877. The 5-by-36-inch board employed metal plugs to switch the circuit from alarm instruments to primitive "box" telephones. Shown here are six metallic blocks—five for subscribers and one for the central office. By inserting and reinserting the metal plugs, the individual lines could be connected for conversation.

Model of the 1878 New Haven switchboard, credited as the first commercial exchange. This board was derived from "dial" telegraph switchboards, used in Connecticut from the 1850s. Each of eight subscribers' lines is connected to one of the metallic studs in the dials, below which are eight "annunciator" switches, each connected to a subscriber's line. The annunciator switches were used to complete and break the annunciator circuits (*above right*) via contact with the long horizontal metal bar, to which were also wired an operator's telephone, a battery, and a ground wire. A third row of switches (*bottom*) was used for signaling.

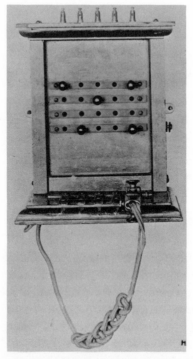

The five-line "plug" switchboard made by Charles Williams, Jr., was the simplest of the manufacturers' switchboards based on similar devices used in telegraph exchanges. The plug board employed a series of crisscrossing vertical and horizontal bars, or metal strips, to make connections. The vertical bars (not shown) were connected to subscribers' lines. The horizontal bars were then placed in electrical contact with the vertical bars by means of plugs inserted in the horizontal strip holes. The circuit thus ran through a vertical strip down to a horizontal strip (connected by means of a plug) and over to a second desired vertical strip (connected to the horizontal strip by means of a second plug).

CHAPTER 2 🐦

Manufacturing Arrangements of the Bell Company to 1879

ON 1 MAY 1877 THE BELL PATENT ASSOCIATION offered its first pair of telephones for lease to Boston banker Roswell Downer, whose pleasure with the speed and efficiency of the instruments helped stimulate a flood of orders for what until that time had been merely an experimental novelty. "We can certainly indorse the telephone as a success," Downer reported, "and believe it can be used practically with many advantages over the old system of telegraphy."[1] The electrical shop of Charles Williams, Jr., where Alexander Graham Bell's assistant, Thomas Watson, had been working on a viable design for the instrument for nearly two years, moved swiftly into production in accordance with specifications provided by the patentees. The manufacture of telephones for commercial use was thus under way.

The market for telephony was narrowly defined, limited to the short-distance transmission of voice between two fixed points. Technically, the early commercial telephone instruments had a transmission capacity of only a few miles and required, according to Watson, "a voice with the carrying capacity of a steam calliope" to be practically audible.[2] Nevertheless, the new business experienced rapid growth. By the end of June some 234 telephones were in service. The number of active telephones would triple by August, quadruple by November, and redouble by the end of the year.

By 1 January 1879 Williams had produced more than 17,500 Bell Company telephones, as well as the auxiliary call bells, switches, and other nonpatented materials (see appendix A).[3] But even this impressive output left the electrical shop far short of satisfying the growing demand.

· I ·

Initial interest in the telephone had been aroused by its sheer novelty as an electrical toy, but by mid-1877, soon after its commercial introduction, the "speaking telephone" was becoming appreciated as an efficient "private line" substitute for the telegraph, which required the presence of a skilled key operator. Even though these first telephones were limited to short-distance communication between two points, there were more than five thousand rentals by early 1878. Then the development of a rudimentary telephone switchboard exchange greatly broadened the dimensions of the new technology. As Watson explained it to one licensee, "This system is simple, any station can be called by the central office and any two put into direct communication with one another." The switchboard overcame the theretofore insurmountable cost and technical barriers to linking more than a handful of subscribers together. The potential for putting a single subscriber "in instant communication with the grocer, butcher, baker . . . and other places and persons too numerous to mention" had particular appeal in larger urban places, where telegraph and messenger services were already a familiar part of business life.[4]

The cost of the telephone to subscribers was substantial for the time. The Bell Patent Association's first advertising circular quoted "terms for leasing two telephones for social purposes" at $20 a year and "for business purposes" at $40 a year. If lines had to be constructed for subscribers, prices for that could "vary from $100 to $150 a mile."[5] Aside from wealthy householders, few private citizens were going to be in the market for telephones. The major demand would come from business, which was already bearing the cost of local telegraph service and paying the wages of skilled operators. This configuration of the market—urban, business, and luxury—would remain much the same until the twentieth century.[6]

The potential market was nonetheless vast, since urbanization and economic growth were proceeding at a rapid pace throughout the United States. This potential market was apparently perceived by the many local entrepreneurs who sought from the Bell patentees licenses allowing them to retail telephone service to the public. The patentees responded by licensing local agents on a geographically exclusive basis to provide telephones (and service) to end users. By July 1878 there were some 16 Bell licensees (with exclusive territorial franchises), operating mainly in New England and the Atlantic Coast states. A year and a half later the market had broadened geographically and there were 172 agencies, with the majority located in the Midwest and the South (see appendix B).

Finding the means to finance and manage telephony's growth posed serious problems for the Bell patentees. They had very little capital to draw on and few sources of support to which they could turn. They could not afford to invest directly in the distribution of local telephone service.[7] They were a small group of entrepreneurs and could not easily administer a process of too rapid expansion. The company's organizational and managerial capacity for marketing the telephone was limited virtually to Gardiner Greene Hubbard's part-time devotion to the business as its trustee.

More than anyone else, Gardiner Hubbard deserves credit as the founder of the telephone business. A man of many parts, Hubbard had made a career as a patent attorney after graduating from Dartmouth in 1841. He was better known as a promoter of new technologies, public works, and education for the deaf, this last interest growing out of concern for his own deaf daughter, Mabel. In the 1850s Hubbard organized businesses to bring a waterworks and gas lighting into his hometown, Cambridge, Massachusetts. There, he also built one of the earliest streetcar lines in the country. In the 1860s he became interested in the problems of communication via telegraphy and the postal service, on which he produced a number of important public policy studies. In 1876 he won an appointment to the federal Railway Commission.[8]

In the meantime, it was through his daughter that Hubbard became interested in Alexander Graham Bell, a young elocutionist who began as Mabel's speech tutor and wound up as her husband. Fascinated by Bell's amateur but serious attempts at telegraph "improvements," Hubbard agreed to back his future son-in-law's experiments. To this end, Hubbard joined with Bell and Thomas Sanders, a wealthy horse breeder and leather merchant and the father of another of Bell's deaf pupils, to form the Bell Patent Association in early 1875.[9] When Bell's experiments led to the discovery of the telephone in 1876, it was Hubbard who worked hardest to bring the two patents on it to market.[10]

At first, Hubbard tried to sell Bell's invention to the Western Union Telegraph Company, to no avail. Having failed to interest the only plausible buyer in their patents, the members of the Bell Patent Association were compelled to establish the value of the telephone themselves if they were to reap any financial rewards. Alexander Graham Bell traveled widely to demonstrate and lecture on the technical feasibility of the telephone, but he had no taste for business. Sanders, who had enough faith in the telephone that he would invest over $100,000 of his money in it before earning any return, struggled from his tannery in Haverhill to keep track of the company's finances in Boston. It was left to Hubbard, who had no financial acumen but plenty of promotional zeal, to establish the means for producing and distributing the new technology.[11] Amid his labors as railway commissioner, Hubbard negotiated terms with local agents, who, in return for a patent license to distribute the telephone, agreed to construct the transmission lines and place the telephone into service.

Hubbard was successful enough in finding agents to get the business started. Rapid expansion of the business, however, was tightly circumscribed by the very limited ability of the patentees to finance and manage it directly and by Hubbard's limited ability to find others to do it. The entire venture therefore depended upon control of the patents. Without them the patentees could quickly be driven out of business by anyone who could finance and manage growth more rapidly. Even with the patents, their position was tenuous because their telephone could be easily copied and sold by others. In a nation as large and populous as the United States in the 1870s, it could be difficult (and very expensive) to hunt out and sue all of those who might infringe on the Bell patents.[12]

These considerations dictated a cautious and costly approach to the distribution of the telephone. While the patentees were eager to develop the market, they were determined to retain tight control over the patented instruments. Telephones were to be leased, not sold (see appendix C), to licensed agents, who serviced subscribers. In that way, the Bell interests could draw on the labor of their licensees, who also absorbed the capital expense associated with line construction. But under this arrangement, the patentees had to absorb the capital expense of the instruments. This policy protected the legal rights of the Bell interests, and in the long run it would assure a continuing stream of revenues; but in the short run it served only to aggravate a chronic shortage of capital. As the always anxious company treasurer, Thomas Sanders, eventually discovered, the company's assets were very soft. Accounts receivable were not always collectible, and the patents themselves (which he believed were widely pirated) were not yet creditworthy.[13]

Nonetheless, protecting the company's patent rights remained the paramount concern of the founders, and leasing the telephones afforded the means by which the company could keep track of the instruments in use. This concern reached back to the supply side of the business in rigorous fashion. Development, production, and ultimate responsibility for repairs were vested in a single producer under the close scrutiny of Watson, the Bell Company's first superintendent of manufacturing. Confining telephone production to the physically proximate shop of Charles Williams, Jr., allowed the patentees to direct carefully the distribution of telephones under the leasing policy and to shepherd further development of the still primitive transmitter and receiver.

Given the well-entrenched acceptance of the telegraph, the qualitative development of telephone hardware was a major concern of the patent holders from the beginning. In its initial private-line application, the telephone's competitive advantage over the telegraph was its potential for instantaneously interactive, two-way communication without intermediaries or procedural delays. Therefore, the market value of the telephone was a function of its proven reliability for transmitting and receiving clear conversation. So worried was the Patent Association's trustee, Gardiner Hubbard, about the technical reputation of the telephone in its early commercial phases that he firmly ordered Watson to "send out no instrument until it has been thoroughly tested." Each subscriber was assured that the "instruments will be kept in good working order by the lessors, free of expense. . . ."[14] These fundamental policies involving the production and distribution of telephones were continued after 9 July 1877, when the first Bell Telephone Company came into existence, replacing the less formal trusteeship of the Bell Patent Association.

· II ·

From 1877 until the spring of 1879, the Bell Company relied exclusively on Charles Williams, Jr., for the production of its telephones and for almost all its telephone apparatus. The choice of Williams's shop had been easy to make. Charles Williams, Jr., was Thomas Watson's employer, and his shop on Court Street in Boston had provided hospitality for Bell's experiments in 1875. There Watson had nurtured the development of Alexander Graham Bell's primitive instrument until it had become sufficiently reliable to be placed on the market. When the Bell Telephone Company opened its office next-door, the proximity assured the harmony of relationships and close communication deemed necessary to protect the patents while working on the technical development of the instruments.

Williams, moreover, was willing to risk the uncertain financial prospects of the fledgling telephone enterprise and was forbearing with his credit. His "faith in us and in the future of the telephone," Watson remembered years later, "was so great that he strained his resources and credit to the limit not only by enlarging his shop . . . but also . . . by giving us plenty of time to pay our big bills."[15] The arrangement was as expedient as it was natural. It was very much in keeping with the business relationship common to the mid-nineteenth century: it was personal, informal, and essentially unstructured in any bureaucratic sense. Not even a contract was signed between the Bell Company and Williams until mid-1878, although it is clear that an understanding existed between the parties, as did agreements giving Williams virtually exclusive manufacturing rights for the telephone and telephone apparatus.

Equally characteristic of the business setting of that time was the existence of one exception to this agreement—a rather odd arrangement for manufacturing rights with a Michigan licensee (see appendix D). In the highly personalized business system of the nineteenth century there was at least one exception to every rule, and at times there might not even be a rule. Little information exists on the way in which the Williams shop functioned on a day-to-day basis. It appears to have been a large but typical machine shop of the period, staffed by men who were more craftsmen than laborers. Its novelty lay in its production of electrical apparatus at a time when the industry was not yet even a category in the United States Census of Manufacturers. In 1872, when Watson went to work there, the shop employed twenty-five men producing telegraph apparatus and fire alarms "in small quantities." It was alongside these machinists that the eighteen-year-old Watson served his apprenticeship in this "interesting and manly trade." While cultivating his "love for expressive speech" in profane exchanges with his bench partner, Watson mastered his new calling and experimented on the side with various engines.[16]

An inveterate tinkerer, Watson felt at home in a place that was a haven for those inventors who could find financial backing for their experiments. Alexander Graham Bell found in Williams's shop a congenial laboratory for pursuing his interests in electrical telegraphy. Having become fascinated by communication via the "occult force of electricity," Watson befriended Bell and helped him bring the idea of voice-grade transmission over wire to practical fruition. In the summer of 1876 Gardiner Hubbard, Bell's financial "angel," hired Watson away from Williams for a 10 percent interest in a "telephone" patent. When commercial telephone production began in 1877, Watson, who previously had made practically all the telephones in existence "with my own hand," returned to the Williams shop. He supervised the construction of the telephones, trying all the while "to improve them and reduce their cost."[17]

In this early phase of the business, the technology, although simple in principle, required considerable development. Initially, the telephone was an awkward affair contained in a bulky rectangular box. Under Watson's guidance, and with the help of an independent group of scientists at Brown University, the Williams shop brought the telephone instrument (transmitter and receiver) through no less than four distinct stages of development in shape and materials between May and August 1877. Gradually, they made the telephone more technically reliable and convenient. Watson also worked on the development of signaling devices; these underwent several modifications until in a year's time a polarized call bell, which became the basis for all signaling well into the twentieth century, was fashioned and patented. From an independent inventor the Bell Company was able to purchase a patent to an automatic switch that would change the telephone from the standby to the talking condition. By 1878 Charles Williams, Jr., was producing all these important pieces of equipment essential to each subscriber for telephone service.[18]

In January 1878, however, the industry's technology experienced a fundamental transition. The demonstration of a commercially feasible switchboard by a Bell licensee in New Haven, Connecticut, promised to refocus the industry's technical development, shifting attention from subscriber station equipment to central exchange apparatus. The emergence of an interconnected system of telephone lines in which any one subscriber could be placed in communication with any other portended not only great opportunities for the growth of telephony but also significant managerial and technical problems for the business. The licensees would have to become operators of a dynamic and continuing service—the ongoing connection and disconnection of subscribers—while the Bell Company would be under increasing pressure to coordinate intelligence on best practice and the development of hardware.

The problem was at least contained by the Bell Company through its control of the telephone patents. The company could not hope to control the development of switching as directly as it tried to control the telephone. Switchboard technology was highly variegated, even at this early date, and was subject to frequent change; in the absence of a controlling patent on the principle of switching (which had its roots in telegraphy), the exchange field was also wide open to competition.[19] Manufacturers were free to produce an array of switchboards, as well as wire, office fixtures, line materials, and tools—all of which were as essential to telephony as they were to telegraphy. Still, to obtain their telephones, they had to turn to Charles Williams, Jr.

While Williams himself began manufacturing switchboards for sale to licensees in July 1878,[20] his relationship to the Bell Company remained centered on the manufacture of telephones and call bells. When the two firms met on 1 August 1878 to sign a formal contract defining their relationship,[21] the agreement centered on subscriber station equipment. It specified the terms under which this equipment was to be inspected, priced, and distributed. The Bell Company control over telephones was the main issue. In the three-page, handwritten contract, the Bell Telephone Company agreed to purchase all of its telephones directly from Williams, paying $1.60 for each hand telephone and $2.45 for each box telephone. Each purchase was subject to the inspection and acceptance of the company's superintendent (Watson). To protect its patents, the Bell Company was determined to keep close control over the shipping of telephones (which Williams carefully numbered in series) and was unwilling to allow any transactions to take place directly between the agents and the manufacturer. Watson himself shipped all of the instruments.

Call bells were a different matter. While they were essential to the provision of telephone service, the patents on them did not control the principle of signaling in the same way that the Bell patents controlled the principle of telephony. Nor, because of the costs involved, did the Bell Company care to keep in its hands the business of purchasing, shipping, and repairing the call bells.[22] These Williams was required to ship himself, with no guarantees of purchase or lease. In the contract, Williams agreed to manufacture call bells in return for protection against any and all suits for patent infringement. He was to sell the bells for eight dollars or to lease them for an annual rental of five dollars (in those special cases where the Bell Company had agreed to pay a commission on rentals). Williams was entitled to draw upon the consignee at a rate of three dollars for each bell shipped, but in the normal course of business he was not guaranteed payment for call bells. Nevertheless, the manufacture of bells, like that of telephones, was subject to the inspection and approval of the superintendent.

To avoid becoming too dependent on its source of supply, the Bell Company reserved the right to terminate this agreement with ninety days' notice should Williams fail to furnish equipment promptly and satisfactorily. In turn, the company agreed "to use their best endeavors to induce their Agents to purchase the Supplies needed in their business" from Williams (including supplies other than telephones and bells). Williams would be indemnified against losses incurred as a result of inventory overstocks due to technological change.

In short, the contract, though not particularly restrictive or elaborate, embodied principles that would long have an impact on the Bell Company and the industry. These principles included insistence on the manufacturer's commitment to the prompt production and controlled distribution of high-quality telephone equipment at prices fixed by the Bell Company. These principles, which had already been developed informally by the time the contract was executed, were born out of the company's need to assure itself a reliable source of supply while, above all, keeping very tight reins on the telephone patents. Thus by mid-1878 the patent holders, through their control of the supply and distribution of subscriber station equipment, hoped to keep control of a growing business—a business conducted by agents, not employees, who were managing switching systems, not merely distributing hardware.

The arrangement with Williams, however, soon fell short in one critical respect. Although it enabled the Bell Company to monitor the quality and distribution of patented equipment, it did not satisfy those licensed agents who wanted to keep pace with the growing demand for telephone service. Complaints of slow deliveries appeared as early as October 1877, and Watson, who was made responsible for filling orders, responded by favoring the sources of greatest pressure. "There seems to be neither order nor system," observed Hubbard in November; he instructed Watson to fill orders in precise order of receipt "without any deviation." Already there existed a backlog of from two to three weeks, and as Hubbard explained, "It will not answer any longer to divide the day's product among the most urgent. . . . Judging from the past, there is no reason to think that you will catch up to your orders in about two weeks [as Watson had hoped]: as the orders received are daily growing upon the orders filled."[23] Williams, meanwhile, was trying to expand his shop's capacity. In August 1877 he promised to increase production from twenty-five telephones daily to fifty.[24] Although actual production mounted steadily over the ensuing months,[25] he incurred significant expenses as he tried, in vain, to stay abreast of demand.

The additional expenses incurred by Williams in his attempt to increase capacity were significant because the Bell Company was chronically short of capital and had from the beginning drawn heavily upon the credit extended by Williams. This necessarily limited Williams's ability to step up output. By January 1878 he required a twenty-five-hundred-dollar advance paid to him out of pocket by Bell treasurer Thomas Sanders, who worried that the company would soon be unable to meet its growing obligations to the manufacturer. Williams complained incessantly about late payments as the Bell Company fell into chronic arrears. Sanders and Hubbard in turn were chagrined by Williams's failure to provide exact accounts of his costs, but at bottom they remained grateful for his good will and extensions of credit.[26]

Without Williams's sometimes grudging willingness to tolerate tardy payment of his bills,[27] the Bell Company would have been hard-pressed to promote the expansion of the industry. Williams, of course, was gambling on the ultimate financial success of a new technology in which he had already invested a great deal of time and expense and for which he had exclusive manufacturing rights. The Bell patent holders, for their part, could do little about their chronic shortage of capital. Their cash flow was subject to the vagaries of their accounts receivable, which were dependent on the ability or willingness of the operating agencies to forward their telephone royalty payments on time. Bell had no effective means for managing licensee relations on a day-to-day basis, and in the absence of strict accounting controls, the company could do little more than wait for the operating agencies to pay. In the meantime, the firm leaned heavily on Williams's forbearance.

There were other advantages to working exclusively with the Williams shop, advantages relating to the company's need to innovate while keeping its patented technology under close control. In the first place, the shop was assiduously attentive to the problems of development. Through the personal link provided by Thomas Watson, Charles Williams's machinists became a responsive technical corps for the Bell patent holders, enabling changes in the design of telephone and signaling apparatus to go forward apace. Production volume no doubt took a back seat to tinkering and fashioning in a shop whose machinists were more craftsmen than specialized production workers, but to a great degree this was a desirable trade-off. A second benefit was the close control that could be maintained over instruments manufactured under the patents, a concern made explicit in the manufacturing contract of August 1878. Since the technical barriers to entry into the telephone business were slight, monitoring of the movement and location of each telephone instrument was seen as the key to enforcement of the integrity of the Bell patents (see appendix C). In effect, the Bell Company decided that these advantages outweighed the need to answer all of the demands of the operating agencies. If it had been in a stronger financial condition, it might have achieved all of these objectives, but in the real world of the late nineteenth century, the company found it necessary to compromise in order to shore up its patent position.

· III ·

The Bell Company's patent position became even more tenuous in early 1878, when the world's largest corporation, the Western Union Telegraph Company, entered the telephone business with its own patents pending. Once the commercial viability of the telephone had been established, it was natural that Western Union would want to reap the benefits of a technology that could be applied to at least one small segment of the communications market, the private-line business. By the time Western Union entered the field, moreover, it was becoming apparent that the telephone had even broader applications and that the Bell Company was making a foray into markets near to the heart of the telegraph business.

It was around the emergent local switchboard exchange business that the battle lines formed. In February 1878 the Bell Company published instructions to its agents urging them openly "to use their best efforts for the introduction of the Telephone into the

District Telegraph System." Bell wanted to move into those central urban telegraph offices that dealt with the multipoint interchange of messages via telegraph or messenger either on a local basis or in conjunction with Western Union's long-distance wire network. Here the "advantages of the telephone over the District System are apparent," claimed the Bell Company. "The message is sent through the Telephone to the (central office), its receipt acknowledged, thus saving much time and expense. . . ." The use of telephones would reduce the dependency on skilled telegraphers and would allow parties to carry on conversations.[28] The development of physically interconnected, switched telephone exchanges was also encouraged. "Where the District Telegraph System has not been introduced," advised the instructions to agents, "a District Telephone Company should be organized . . . running from the Central Office to various parts of the city." The threat that these overtures offered to various applications of the telegraph on the local level was not lost on Western Union. It squared off with Bell in a sharp competitive battle that would last over a year and a half.[29]

Western Union was a formidable rival, well armed with its existing wire plant, offices in place (either owned or in other ways affiliated), and vast financial resources. The company, which had long since consolidated most of the nation's important telegraph operations, had well over forty million dollars in capitalization. It enjoyed three million dollars a year in net profits from its busy nationwide network of wires and offices in nearly every city, town, and established settlement in the United States. Western Union had little to fear, it seems, from a small contingent of modestly capitalized entrepreneurs whose claim to a controlling patent could be stalled in court at a cost insignificant to a corporate giant.[30]

Western Union's array of telephone patents was impressive. It owned the patent claim of Elisha Gray, whose caveat on the principle of the telephone had been filed within hours of Bell's first patent in 1876. To this it would soon add the claims of Amos Dolbear (whose patent application of September 1876 contained improvements thought by many to be controlling) and Thomas Edison (who had devised a carbon-button, variable-resistance transmitter, different in principle and technically superior to Bell's liquid acid and magneto devices). Telephones made under these claims were distributed through the Gold and Stock Telegraph Company, a Western Union subsidiary that operated under license from the American Speaking Telephone Company, a firm that Western Union had created to hold its telephone patents.[31]

The Bell interests' initial response to the prospect of a frontal assault on its patent monopoly was cautious. The company's New York correspondent, Charles Cheever, warned in December 1877 that any legal challenge to Western Union would become an "extremely tedious" process "lasting over several years." Convinced that Western Union was "determined to crush us by fair means or foul," Thomas Sanders, who had sunk more than one hundred thousand dollars of his own money into the development of the Bell Telephone, tried to persuade Hubbard to "make the best terms we can with this powerful combination," even if it meant selling the business.[32]

Sanders, in fact, was feeling quite desperate. The onslaught of Western Union exacerbated a growing tension between the Bell Company's treasurer and its president, who feuded throughout the spring of 1878 over the company's precarious financial position. Hubbard apparently spent freely and used what credit the company had to run up a debt of twenty-five thousand dollars by the end of February. In letter after letter, Sanders scolded his partner for his "blissful disregard of money" and his seeming obliviousness to the various pressures that were threatening to send the firm into bankruptcy. Western Union's entry was scaring off the company's creditors. Worse yet, "our agents are paralysed. . . . Williams calls daily for thousands, and if in addition to that we have to pay lawyers, where will the money come from?" Sanders wrote that he was paying Williams out of pocket and had been forced to lie to at least one agent about the company's condition so "that he might not think we are bankrupt." He thought that "we may and I trust will laugh over this one day, but it is a sorry matter now. If I break down under it, deal gently with the widow and the fatherless!"[33]

Hubbard was serene by comparison. In the face of the threat from Western Union he maintained a confident posture. He did agree with Sanders to consider some form of compromise with Western Union, but whereas Sanders was willing for the Bell interests to settle for a quarter of the stock in a new, Western Union–controlled telephone company, Hubbard suggested the creation of a jointly owned company in which the Bell interests would hold a half-interest and he would serve as president. Because the Bell interests were so weak financially, however, any hope for a merger on these terms was inconceivable to Western Union.[34] Moreover, the telegraph company's president, William Orton, had long harbored a grudge against Hubbard, who a few years earlier had lobbied intensely (albeit unsuccessfully) for a law to create a federally chartered "Postal Telegraph" service. Orton, who in the winter of 1876/77 had disdainfully refused Hubbard's offer of the Bell telephone patents for one hundred thousand dollars, apparently relished the thought of a corporate fight to the death.[35]

The president of Western Union was denied that simple pleasure when he died in April 1878, but his firm pressed ahead with the competition. Equipped with the Edison transmitter, introduced commercially in June, it was now in a stronger position technologically to achieve victory in the struggle. Western Union stepped up the pace of what had become a race to secure the most potentially lucrative urban exchange markets. It cut prices (and, on occasion, Bell lines) and began to buy out Bell agents. By offering potential agents (often the owners of local district telegraph companies) access to already constructed lines, lenient terms of credit, indemnification against losses in patent suits, and lower-priced telephone equipment (including a superior transmitter), Western Union succeeded in acquiring a strong foothold in telephone communications by the year's end. By 1879, the telegraph company had prevailed in several important urban markets, including Chicago and New York, and its rate of growth now exceeded that of the Bell Company.[36]

· IV ·

In open business competition, as in war, the mastery of supplies is crucial to success. To support its foray into telephones, Western Union could marshal the productive capacity of the Western Electric Manufacturing Company of Illinois. By all accounts, Western Electric was the largest electrical apparatus producer in the country. It had been controlled by Western Union since 1872, when the

telegraph company and one of its senior officers had together acquired more than two-thirds of the manufacturer's capital stock. By
1879 Western Electric had become a significant asset, providing a
large force of experienced electrical machinists and a substantial
capacity.[37] Since Western Union was integrated vertically into
production on a more formal basis than was Bell, and since it owned
stock in its manufacturer, it stood to share in the profits from the
production as well as the leasing of telephone equipment. In that
sense, too, the telegraph company was well armed for a competitive
struggle over this new branch of the communications industry.

As mentioned above, Western Union owned stock in its
manufacturing firm; but the relationship between the two businesses
was more personal—in the nineteenth-century style—than this
situation suggests. To understand this relationship, it is necessary to
backtrack a bit. In 1856 the various telegraph manufacturing shops
doing business with Western Union were consolidated into two main
plants, at Cleveland, Ohio, and Ottawa, Illinois. In that year
Western Union acquired the latter; the Cleveland plant was purchased by George Shawk, a former telegraph repairman. In 1869
Shawk sold part of his interest in what was then a modest-sized shop
to Enos Barton, a young telegrapher from Rochester, who had
borrowed money from a friend and had mortgaged his small property
holdings in order to go into business for himself. The inventor Elisha
Gray (who later claimed to have invented the telephone) bought
Shawk's holdings later that year and established the firm of Gray and
Barton. By the year's end, Gray and Barton had brought in a third
partner. General Anson Stager, former chief communications offi-
cer of the Union army and now a vice president of Western Union,
contributed $2,500 in capital for a third of the business and per-
suaded Barton and Gray to move to Chicago.[38]

Gray and Barton further linked their future to that of West-
ern Union's in 1872, when they agreed to the incorporation of the
Western Electric Manufacturing Company, capitalized at a healthy
$150,000. Stager had persuaded William Orton, president of West-
ern Union, to take one-third of the stock in the new company, and
Stager himself took a third. The remainder was held by friends of
Stager's, by employees of Western Electric, and in small measure by
Gray and Barton. Three of the five directors were also directors of

Western Union, which then closed its Ottawa shop in anticipation that the Chicago factory would supply most of its needs. The Chicago plant was able to meet these demands until 1878, when the company found it necessary to open a branch in New York City.[39]

For their part, Gray and Barton, now small shareholders, were placed on salary as electrician and secretary (i.e., general manager), respectively, of Western Electric. They shared handsomely in the prosperity of the firm. From 1872 to 1874, in the midst of the great depression of the 1870s, Western Electric's employees nearly quadrupled in number to one hundred, already twice the size that Charles Williams, Jr.'s shop would reach four years later.[40] By 1878 the Western Electric Manufacturing Company, largely by virtue of its association with Western Union, was the largest producer of electrical apparatus in the world. In addition to telegraph equipment, signal boxes, registers, annunciators, and fire alarms (which Gray and Barton had been making before 1872), Western Electric was producing a large variety of other electrical equipment. Its products included sophisticated measuring devices (mirror galvanometers, electrometers, rheostats), condensers, railroad instruments, arc lamps, mimeograph pens, and high-tension arc machines.[41]

With its broad experience in almost every contemporary application of electricity, Western Electric had no trouble gearing up quickly to produce an array of telephone appliances in large quantities. By 1879 it had become especially adept in large switchboard technology, and the quality of Western Electric equipment, according to Watson, was high enough to earn the admiration of Charles Williams's craftsmen.[42]

· V ·

Prior to Western Union's entry into the telephone business, Western Electric had actually been serving as an agency for the Bell interests in Chicago. According to some stories, Anson Stager, as Western Union vice president, had recommended against the purchase of the Bell patent when Hubbard had offered it to the telegraph company in late 1876. Once the telephone's potential for profit had been demonstrated, however, Stager became eager to hop on the bandwagon. In his capacity as president of Western Electric, he sought and received a license from the Bell Company to develop private-line telephone service. At the same time, Enos Barton sought permission to manufacture telephone equipment, a request that was apparently refused by the Bell Company. The available correspondence indicates that by year's end, Western Electric was building a steady Bell rental business in Chicago.[43]

Soon, however, relations between the two companies began to sour. As early as October, Hubbard began to suspect that the business in Chicago was not being pursued in earnest, a charge to which Barton responded (somewhat disingenuously) by saying that Western Electric was expending efforts on the telephone to the detriment of its own business in the printing telegraph. He also cited technical difficulties owing to transmission interference.[44] But Hubbard's suspicions persisted and grew, fanned by reports from third parties that Western Electric was striking an agreement to make telephones for Western Union under the patent claims of Elisha Gray. In December Stager admitted that this was the case.

Anson Stager saw no inherent conflict of interest in these developments. Enos Barton felt that Western Electric should terminate its role as Bell agent, because it was preparing to manufacture telephones for the Gold and Stock Company, but Stager insisted that he was still operating in good faith on behalf of the Bell interests. With righteous indignation he bridled at Hubbard's plan "to send an agent [to Chicago] to assist in working up the business." Stager grumbled that this was an unwarranted "reflection upon our business capacity and integrity."[45]

Conflict of interest was not a well-developed notion in nineteenth-century business practice. The ethics of allegiance of man to firm were by modern standards muddy at best. If Stager saw no necessary conflict between Western Electric's role as a captive supplier to Western Union and as an agent for its sole competitor in a new market, it was because he believed in his own ability to wear more than one hat—even in the competitive fray. Equally remarkable by modern standards was Hubbard's sanguinity even after Stager had warned him that "it looks as though competition will be sharp"[46] and that Western Electric would be playing both sides of the fence, as producer for Western Union and distributor for Bell. Hubbard was also unshaken by Thomas Watson's reports from Chicago (while on an inspection tour of Bell agencies in the spring of 1878) that he had seen nothing but Gold and Stock apparatus on the shelves of Western Electric. Watson thought Stager and Barton to be double-dealers, "a set of men capable of almost any game." He concluded that "the W. E. Co. are doing all they can to hurt us."[47] While Hubbard reacted by recruiting a Bell agent from Missouri to go to Chicago to try to improve the company's business there (and to keep an eye on Western Electric), he continued to correspond with Stager, hoping for the best.[48]

In early June 1878, long after it had become apparent that Western Union (with Western Electric's support) was outrunning Bell in Chicago, Hubbard received a proposal from Enos Barton suggesting an end to the competitive rivalry. After all, no reasonable businessman wanted to risk an unprofitable battle for the whole field when he could do well by half. Barton proposed that the American District Telephone Company, Western Union's Chicago agency, split its revenues evenly with the Bell Company and use both Bell and Gold and Stock telephones. "In this manner," explained Barton, "you would be assured of a handsome revenue from the start, and would make more money than you could possibly from a competing organization, for such an organization could not hope to secure one-half the business, on account of the advantage that the American Dist. Tel. Co. already has in the occupation of that field."[49]

Hubbard, perhaps naively, pronounced the offer "fair and liberal," but he noted that it asked for the Bell Company to depart from its usual procedure for renting the telephone. There were other impediments to the offer as well. What Hubbard did not say (but what he surely must have realized) was that the offer, if accepted, would amount to an abandonment of the Bell Company's claim to exclusive patent rights on the telephone.[50] In any case, Watson's animated displeasure with Western Electric was exacerbated by the Chicago agent's reports of Barton's evasiveness and his attempts to maneuver "between two forces." Although the officers of Western Electric and the American District Company might have believed sincerely that a viable cooperative business relationship could be worked out, the conflict of interest was too great. As Charles

Cheever, Bell's New York agent, asked in December 1877, "How will all this new W. U. arrangement of Speaking Telephones work in harmony with the concern who are your interests in Chicago? Can its officers so prominently connected with the W. U., which is now a decided opposition interest to ours, fairly & honestly represent the Bell Telephone, no matter how desirous they might be of so doing[?]"[51] The answer was no. The proposal was dropped, and Western Electric moved openly into unequivocal partnership with Western Union's telephone enterprise. One year later the Bell Company would lose its struggling Chicago agency to Western Union in one of the latter's many successful raids.[52]

· VI ·

It was in this crucible of competition that the telephone became an instrument of general utility for intraurban, mercantile communications. Demand for telephone service increased sharply after the development of local exchange services. By mid-1878 the Bell Company had developed a "district exchange system"—a central office combining telephone, telegraph, and courier services—in every major city in the United States.[53] Each such exchange became a focal point of competition, since Western Union challenged Bell for business in the more heavily populated cities and towns. Competition quickened the pace of development, as agents on both sides came to realize the practical necessity of being the first to "occupy the field" (to use the contemporary term). Thomas Watson, writing from the field in May 1878, found a silver lining in the gathering clouds, insisting that "the W. U. opposition has done and is doing us more good than harm." In St. Louis, for instance, "where our agents greatly bewail the opposition . . . , I venture to say that had it not been for this opposition they would not have developed ½ the business they have done."[54] During the year 1878 the number of instruments in the hands of Bell licensees alone more than trebled (see appendix A), lending credence to Watson's analysis; but the problems that Bell encountered in Chicago and elsewhere confirmed the darker, more perilous aspects of competition.

Facing rising demand, changing technology, and the strategic and tactical problems of competition, Hubbard and Sanders decided to revamp the Bell Company, changing it from a small, nineteenth-century mercantile operation managed by part-time directors into something resembling a modern, professionally managed business corporation. Even while they were pursuing negotiations with Western Union over a possible combination of interests, they were restructuring their business along these lines. They hoped thereby to meet their present and anticipated needs for capital and more effective management.

Yet Bell's potential to respond to competition and demand remained severely constrained by a shortage of capital. When in early 1878 Sanders had been scrambling to cover current liabilities out of his own personal funds, he had recruited a group of investors to participate in the formation of a corporation that would serve as a superagency for New England. The new investors, men from Rhode Island and Massachusetts, along with the Hubbard family, Sanders, and Watson, put up $100,000 for 1,000 shares of stock in the New England Telephone Company (incorporated under the laws of Massachusetts on 12 February 1878). Another $100,000 in stock was paid into the Bell Telephone Company. The new corporation, with Hubbard as trustee, bought its telephones from the Bell Company and then leased them to local agents.[55]

The arrangement had two important benefits consistent with the patent holders' need to raise funds without losing control of their patents. First, it provided a much-needed infusion of capital for the patentees so that they could increase the manufacture of telephone equipment and help in the development of telephone exchanges. The exchange business was especially expensive, because it required extensive investment in wire and switching equipment. (It was the New England Company that financed the first commercial exchange, in New Haven.) Second, by virtue of the interlocking of Bell Company stock and directors, the patentees retained tight control over the business while admitting new investors.[56] The patentees also may have intended for the New England Company to acquire agency properties in the region, but after its purchase of the Boston agency for $3,500 in cash in February or March 1878, it made few other investments of this kind.

For a time this helped keep the wolf away from the door. Charles Williams, Jr., now enjoyed a more reliable schedule of payments for the production of telephone equipment. But financing and managing a growing business under the stress of competition became increasingly problematic. In April it became clear that Bell's New York agency was failing under the dual strain of mismanagement from within and rivalry from Western Union for subscribers. The Bell Telephone Company went to the rescue, buying out the New York agency, with all its outstanding obligations.[57] This solved the situation in New York, but it stretched Bell's own financial resources to the limit and also made it responsible for managing a crucial but unhealthy operation in the nation's major urban and business center.

Bell, however, lacked the administrative resources to cope with either the growth of its own business or the problems of its licensees. With the exception of one full-time office clerk and Thomas Watson (who when he was not traveling spent the bulk of his time at the Williams shop), the Bell Company had no real day-to-day managers before May 1878. To meet the pressures of the growing and competitive business, Sanders closed the leather business in which he had made his life's fortune. He thereafter devoted himself fretfully to the financial problems of the Bell Company.[58] Hubbard, meanwhile, in his peregrinations for a congressional commission on the railroad, had helped out by recruiting a full-time manager for the business. Hubbard had persuaded Theodore Vail to leave his job as superintendent of the Railway Mail Service, where he had successfully developed a system for coordinating mail and telegraph communications, to become the Bell Company's first general manager. Vail's arrival was applauded by Watson, who was "delighted with the prospect of getting rid of the part of my work I didn't like." Now Watson could concentrate fully on technical matters and leave to Vail the correspondence over business arrangements with the operating licensees.[59]

Vail was the son of a struggling ironworks manager and a young cousin of Alfred Vail, a noted inventor who had developed the Morse telegraph into a practical device. After a false start in the apothecary trade, the younger Vail had turned to the wires as a telegrapher for hog brokers in Manhattan in 1864. His fondness for billiards and the night life and his inattention to business had gotten him fired. With his parents, in search of new opportunity, he had migrated to Iowa in 1866. There he had homesteaded and played baseball before landing a job as a telegrapher for the Union Pacific. In 1869, Vail had joined the Railway Mail Service, where he had made his mark as an efficient administrator and reformer and became the agency's superintendent in 1876. It was in this capacity that he met Hubbard, the railway commissioner, who spoke glowingly of opportunities in telephony. At age thirty-three Vail decided to join the Bell enterprise in the wake of a tawdry political squabble over the status of his expense account of five dollars per diem (on top of his annual salary of twenty-five hundred dollars).[60]

Vail was brought in with a significant raise in salary and a promise that the Bell telephone business would boom. For a while, however, he was lucky to get paid at all.[61] His first important task involved a largely unsuccessful attempt to put the failing New York agency on a sound footing.[62] It was fortunate for the Bell Company's future that Vail did not allow that thankless task to deter him; and although he worked out of an office in New York, he had a salutary influence on the loosely managed state of affairs in Boston.

Vail's coming signified not only the beginning of the transformation of the Bell enterprise from a loosely coordinated partnership into a professionally managed corporation but also the passing of Hubbard's personal dominance. Vail's first important contribution to this process came when he helped Sanders to persuade Hubbard that the Bell Company had to be reorganized as a corporation under wider ownership, along the lines of the New England Company.[63] Loath to part with control, Hubbard nevertheless had come to realize that the business could no longer be operated on the basis of current receipts. Reluctantly, he consented to the creation of the Bell Telephone Company as a Massachusetts corporation under an enlarged directorate. Hubbard became president, but he lost the power he had possessed as the old association's trustee; now he had to share decision-making authority with an executive committee made up of himself, Sanders (who had invested far more of his personal fortune into the business than had Hubbard), and George L. Bradley, a prominent stockholder in the New England Company.[64] The firm was in a sense outgrowing the individuals who had promoted its early development. In the years that followed, as the Bell Company expanded and strengthened its operations, highly personal relationships would continue to give way to more formal organizational ties.

Meanwhile, the financial needs of the company continued to mount as even the funds brought into the business by the new stockholders were stretched. The costs of doing business were further increased by the growing volume and complexity of the technological developments taking place. In June 1878, Western Union placed in service a strong carbon-button transmitter based on a Thomas Edison patent (which the Bell Company, despite its need for a more powerful transmitter, had refused to purchase in April). So much better was the Edison transmitter (which employed two solid electrodes in contact with each other) than the erratic and often weak Bell model (which relied on voice-induced, rather than battery-powered, currents) that during the summer and fall of 1878 Western Union gained many new subscribers on the basis of its superior technology. By September Vail acknowledged, "We want a carbon telephone." Feeling this pressure, Hubbard hired Emile Berliner, who had a relevant patent pending, to work on a microphonic transmitter for the Bell Company. As it happened, however, a viable, patentable, and superior microphonic transmitter—powered by a battery and using electrodes consisting of a block of hard carbon and a platinum bead—was offered to the company by Francis Blake, an independent inventor from Weston, Massachusetts. This time the company bought the patent. Clearer in sound and easier to adjust than Edison's, the Blake transmitter soon helped the Bell Company to recover its reputation for technical excellence.[65]

One lesson to be learned from the company's experiences with the Edison and Blake transmitters was that the Bell Company could no longer rely on itself or on a small group of easily identifiable friends for its technical needs.[66] Monitoring work done by its licensees, by competitors, and by independent inventors for useful or potentially threatening patents was likely to be essential to the firm's survival and growth. Bell's need for vigilance and intelligence as well as research and development further affected the firm's requirements for personnel and funds. This was particularly true where the burgeoning technology of the exchange was concerned.

The exchange, however, involved more than worrying about

patents. Technical resources would have to be devoted to the problems of operations as well as to the controlling principles of hardware design. The technology of the exchange offered myriad possibilities for switchboard design, and it was as yet uncertain exactly how the switching function might be adapted to substantial growth in the number of subscribers. Nor was it certain what technical and managerial requirements would arise from the multiple connections of large numbers of telephones and apparatus through central switching mechanisms. Sets of equipment and wires inter-linked in this fashion posed what a later generation of engineers would recognize as a "systems" problem.

As the telephone pioneers groped with this new problem, each exchange became in effect an experimental laboratory. The Bell Company conducted a wide correspondence and continuing round of visits in an attempt to assimilate the intelligence being developed in the exchanges. In the spring of 1878, Watson was "very busy all the time" providing instructions to licensees on technical matters and "working out details—to supply all CO's with uniform instruments." By the end of the year, the company was employing traveling agents on a salaried basis to advise local agencies on their business operations and to scrutinize those operations for purposes of accountability; the agents also became itinerant experts on technical problems and solutions. A former employee of a sewing machine company, Oscar Madden, traveled through New York State in this capacity; by 1879 he had done so well in this job that he was offered a position as general agent for the entire country.[67]

In this and other ways the Bell Company was being drawn inexorably into the affairs of the operating companies. Whatever long-term notions the patent holders may have held about taking equity in local operations, their first moves in this direction were more circumstantial than planned. Competition forced Bell to provide financial aid to some of its exchanges, which was contrary to its general policy of relying on licensees to provide the capital necessary for operating the telephone service. In addition to sus-taining operations in the faltering New York agency (by recruiting backers and managers to develop a competitive exchange), the Bell Company had spent at least forty thousand dollars by early 1879 in its effort to promote the development of an exchange in Chicago in the face of overwhelming competition from Western Union. In Buffalo the Bell Company made a number of concessions—including payment of a special commission for subscriber canvassers and lowered rentals on telephones—to its local licensee, who was struggling in the face of stiff competition. There is some evidence that when the New England Company took financial control over the Boston telephone agency in early 1878, Western Union was mounting a vigorous canvassing effort on Bell's home turf.[68]

The financial constraints on Bell tightened accordingly. By the end of 1878 the company badly needed additional funds. Added to the costs of meeting competition in important urban centers were the mounting charges for the Bell Company's own operations and the increased billings from Charles Williams, Jr. In September 1878, after it became apparent that neither negotiations nor confrontation in the marketplace was redounding to the benefit of the Bell Company, the firm brought an infringement suit against Western Union. This, too, cost money. Beset by stiff competition, short of funds, and anxious about its legal position on the patents, Bell found its prospects clouded with uncertainty.

· VII ·

Thus, within six months of the outbreak of competition, the Bell patent holders were responding to the assaults from Western Union on several fronts: reorganizing their corporate structure to secure more funds; recruiting a general manager to organize and operate the business on a more formal and routine basis; devoting more internal resources to research and development; providing financial aid to critical licensees; and, of course, litigating over the telephone patents. Because competition in a technologically enhanced market accelerated the demand for telephones and widened it geographically, the Bell Company was now compelled to pay close attention not only to the expansion and development of the agencies but also to the increase in production. Achieving greater control over the vertical functions of the business continued, nevertheless, to be constrained by limited administrative and financial resources.

Administratively, the Bell Company was being stretched to its limits. The firm was experiencing serious problems both with its numerous licensees and with its sole manufacturer of telephones. It was being pulled further than it wanted to be (or could afford to be) into the daily business of the licensees. This is evident from the rising volume of correspondence in the General Manager's Letter Books for the second half of 1878. Theodore Vail was in constant communication with operating licensees and traveling agents on issues large and small, on matters financial, technical, and managerial.

Earlier in the year the patentees had begun to license agents for district and exchange operations under new, territorially exclusive contracts.[69] These contracts signified the growing necessity for Bell to become involved in the operations side of the business. Through these explicit, formal agreements, the Bell Company hoped to make local investment in exchange operations more attractive by guaranteeing agents and their backers sole rights to local markets under the Bell patents. Joint responsibility for the provision and maintenance of subscriber service by the Bell Company *and* its licensees was assumed. Commissions to agents, which had varied under the less formal arrangements struck by Hubbard, were now made uniform. Bell promised to shoulder the expense of patent litigation and to continue its responsibility for the repair of telephone instruments. By late 1878 the company was using printed forms for these contracts, and it soon began to include a provision for purchase of the local telephone plant "at actual value" should the agency go out of business for any reason.[70] Although the corporate charters of the New England Company and the Bell Telephone Company did not explicitly permit them to hold stock in other corporations, and although this was not the general practice of the firm, the Bell Company found it necessary in some instances to buy stock in the licensees' businesses. Often this was done out of desperation, as in Chicago, where the company was forced in December to take eighty thousand dollars in fully paid-up stock in order to keep the exchange afloat.[71]

As the year drew to a close, the mounting financial problems of the Bell Company called into question the trend toward more involvement in the affairs of the licensees. Sanders wrote to Vail in November to express his distress that the company might have strayed too far "outside its legitimate designated business viz the renting of telephones." This sentiment was seconded by George Bradley, who wanted to conserve the company's resources for the defrayal of lawyers' fees and operating expenses. Sanders, however, was feeling more optimistic than he had been in the spring. He was now willing to place his bets on the patent infringement suit. "The harvest will be plenty enough," he felt, "when our patents are decided in our favor." Meanwhile, "if we use our weapons of defense for extending the business we are lost."[72] Vail and Hubbard believed otherwise, and they argued for increased integration forward into the business of the operating licensees. Without additional capital, however, there was no hope that Bell could continue along these lines and successfully master the competition in crucial markets such as those in Chicago and New York.

· VIII ·

The problems with manufacturing were of a different sort. The Bell Company's relationship with the Charles Williams shop had not changed in principle, although the August contract between the two firms[73] may be seen as part of the overall process of placing the business on a more formal, less personal basis. Actually, the ties that linked Williams to the patent holders had become even closer, in a business sense, when he had purchased stock in the new Bell Telephone Company in the summer of 1878. This, however, had not kept serious problems from developing in the production end of the business. Williams suffered throughout the year from the Bell Company's inability to pay its bills on time, although he "studiously endeavored to keep up the credit of the [Bell] Company." Sanders appreciated this, and in March he dissuaded Hubbard from granting a manufacturing license to parties in New York in exchange for payment of the firm's manufacturing debts.[74]

Still, Williams's exclusive right to manufacture Bell telephone equipment was a mixed blessing. While the Bell interests strove to meet the competition from Western Union and the growing demand for telephones, Williams fell behind in his output. As demand increased, Williams added more workers and produced more hardware.[75] But he could not keep up with the agencies' mounting cries for equipment. In the spring of 1878 the shortage of telephones was somewhat alleviated when the directors of the New England Company decided to abandon Hubbard's policy of requiring two telephones at each subscriber station, one for listening and one for speaking.[76] Soon, however, Williams was again falling behind the orders from the field.

Toward the middle of the year, the competition for the telephone business in the more important metropolitan areas compelled the Bell interests to enlarge their scope of operations. They now had to move into the field on many fronts and more rapidly than they might have if left alone. The rising demand soon outstripped Williams's ability to manufacture equipment. When Hubbard suggested the licensing of other manufacturers in March, however, Sanders countered by supporting Williams's contention that he could manufacture telephone instruments better and more cheaply than anyone else. Williams's employees, wrote Sanders, "understand the manufacture of telephones better than anyone else in the world" and under Watson's guidance made instruments "in a manner which outstrips all competition." "This in a large measure," he insisted, "has been the reason we have so successfully defied the W. U.—the superiority of our telephones."[77] When Western Union cut the prices of its Western Electric–made telephones, the Bell Company tried to counter this thrust by stressing the quality of its instruments (along with more reliable service under patents that ultimately proved to be controlling).[78]

But the pressure on the Williams shop mounted, and the Bell Company soon had even more reason to look for new means of obtaining its equipment. For one thing, the firm's bad experience with the Edison transmitter encouraged it to develop a much broader program of research and development. No longer could it depend solely on the efforts of Watson, who was also charged with the burdensome responsibility of supervising production.[79] From the West, too, came new complaints about the slow delivery of telephones to agents. Subsequently, Vail asked Williams to fill his back orders and to develop an inventory, but this proved to be a vain request. By the end of the year, complaints from the operating licensees were putting more and more pressure on Bell. "Williams' tardiness . . . is getting to be intolerable," G. F. Durant wrote from St. Louis; Grant felt that he had "been often seriously embarrassed by [the] delay." Durant shared the view of many of the western agents, who believed that more reliable suppliers could be found closer to home. Even in New York State, E. J. Hall, of the Buffalo exchange, complained about delays in instruments, switchboards, and office fixtures critical to the development of his business.[80] A single electrical shop in Boston, it seemed, could no longer handle a business that was now reaching national proportions.

By 1879 a new and serious complaint about the quality of Williams's equipment surfaced. This was alarming because one of the major strengths of the shop had always been the high quality of its products. At least two important agents reported Williams's call bells to be less durable (though more powerful) than those of earlier shipments, which, they feared, was causing a decline in subscriber confidence.[81] For the Bell Company, obligated as it was to handle the repairs of its leased instruments, this was a critical problem.

Williams responded to these several complaints by expanding his facilities. By December he had a working force of from fifty to sixty people, up thirteen from June, and he had added two thousand dollars' worth of machinery. Some of his employees were laboring eleven hours a day, and he began to negotiate for additional space.[82] His problems were compounded by the technical challenge of retooling half his telephone production capacity from the Bell "butter-stamp" converters, which doubled as transmitters and receivers, to the Blake transmitter, which required a radically new design, as well as new components. He claimed to be filling his orders promptly, with some lag in switchboards, but by 7 February 1879 he was still shipping only thirty-five telephones per day. He felt that orders were being filled "in due season."[83] But this was not good enough for the agents or for Vail, who two days later yielded to the cries from the field by proposing a substantive change in the company's manufacturing arrangements.

CHAPTER 3 🖛

Manufacturing under the License Contracts, 1879–1881

FOR THE BELL TELEPHONE INTERESTS the turn of 1879 portended a bleak year. Their company, according to Alexander Graham Bell, was on "the verge of bankruptcy."[1] Competition, especially in the major urban markets, was draining the Bell Company's dwindling funds. From the beginning of the telephone enterprise a shortage of capital had influenced almost every aspect of the Bell Company's operations and style of organization. Now, despite the firm's expanded resources, the same problem was threatening the life of the company. In addition to coping with its own rising costs of administration, the firm was now supporting a number of its licensees through discounts on rentals, salaries to agents, and, in some cases, direct subsidies for local operations. The single most serious drain on company resources was the more than forty thousand dollars that the firm had funneled into its Chicago exchange. Legal expenses continued to mount, and none of the original patent holders could afford to sink any more of their own money into the business.[2]

Compounding the basic financial problems of the company was a lack of well-coordinated, full-time management at both the center and the periphery of the business. At the periphery, in the operating agencies, there was a growing need for attentive and aggressive leaders who could develop local business on the competitive fronts. Too many local agents—many of whom had been recruited somewhat haphazardly—lacked the funds necessary for rapid growth and treated the telephone service as a sideline to their other business concerns. All too often, local agents operated on the basis of verbal understandings arrived at with Hubbard, who, as Bell Company trustee, had handled the bulk of the contracting before July 1878. The licensees' bookkeeping was often erratic, and royalty payments were often in arrears. Agents, operating in uncertain, competitive circumstances, lacked timely intelligence on the Bell–Western Union situation (as it moved in and out of negotiations) and were kept in a nervous condition by the apparent disparity in resources and power of the rival patent holders. Many of them also needed help in improving their business and technical practices.[3]

At the center, the Bell Company had problems in handling a swelling body of correspondence from the field, in obtaining reliable accounts and royalty payments, and in assuring its agents that their investments were safe under the Bell patents. Despite the hiring of Vail as the firm's first professional manager in May 1878, over time it became clear to the directors that a number of basic reforms in administrative practice were in order if the business was to be restored to a sound footing.[4] For one thing, the great diversity in the contracts with the operating agents made it difficult for the company to develop appropriate policies for all of them. A higher degree of uniformity seemed desirable as a means of improving performance both at the core and at the outer reaches of the business.

Amid these financial and administrative constraints lurked a difficulty that lay at the very heart of the functional relationship between the Bell Company and its licensees. In the face of rising demand and heavy competition, the patent holders were having increasing difficulty assuring a reliable flow of telephones, apparatus, wire, and switchboards to their exchanges. The Bell Company was faltering in its primary role as supplier of equipment. Bell could hardly demand more of the agents if it could not itself improve the performance of the manufacturing shop of Charles Williams, Jr.[5]

· I ·

During the first half of 1879 the Bell interests took a number of remedial measures to alleviate their financial, administrative, and production woes. To deal with the financial crisis, the original patent holders somewhat reluctantly, but decisively, relinquished control of the firm to new and wealthier stockholders. In December 1878 the tightly held Bell Telephone Company had already made room for two new stockholders on its board of directors. One was Francis Blake, Jr., whose transmitter patent apparently had been purchased in exchange for a large block of stock. The other was William H. Forbes, the scion of the prominent Boston investment house of John Murray Forbes and himself a seasoned financier in shipping, real estate, and railroads. Forbes, in particular, attracted the respect of Alexander Graham Bell, who found him to be a man of "Integrity, Firmness, Good Sense and experience," the only man among the directors "who impresses me as possessing marked ability." When a struggle for power ensued among the Bell Company's directorate in the early weeks of 1879, it was Forbes who emerged as the new leader of a reorganized enterprise, a consolidation of the New England and Bell companies.[6]

The National Bell Telephone Company was created on 17 February 1879 and was incorporated a month later. It succeeded to the rights and property of the New England and Bell companies and broadly expanded the group of investors with a stake in the Bell patents. The new firm was authorized to issue $850,000 in stock at $100 par value, of which 6,500 shares were exchanged for the stock of the older companies. Two thousand shares were set aside for conversion into cash as needed.[7] The new company's office was to be located in Boston. For more than half a year the Bell Company had tried to operate out of New York, but as Vail had discovered, during that time he was isolated from most of the firm's directors.[8] In Boston he could stay in close touch with the firm's financial supporters.

The directorate was reshaped. The patent holders' influence began to wane. Sanders, who had resigned as company treasurer in favor of Bradley (of the old New England Company), was feuding with Hubbard over who was most to blame for the firm's sad state of affairs. Into the breach moved the thirty-eight-year-old Forbes. He was elected president of National Bell, since Hubbard could no longer command the complete trust he had once enjoyed. Sanders and Hubbard were outnumbered on the executive committee, which now included Forbes, Blake, and another newcomer, R. S. Fay.[9]

In Forbes, Vail found a skilled financier and manager to whom he could report and with whom he could work on a business-like basis. Forbes, who in 1859 had disgraced his French immigrant family when he was expelled from Harvard for vandalizing its chapel and assaulting its watchman, had atoned for his transgression through long hours of labor as his father's clerk at J. M. Forbes & Co. After distinguishing himself in the Civil War as a company commander in the Union cavalry and as a prisoner of war, he had returned home to take up a full partnership in the family counting house. Between 1866 and 1878 Forbes had traveled widely to oversee his company's far-flung investments and had corresponded heavily on the myriad details of its varied enterprises.[10] It is his meticulous attention to detail, as well as his solid grasp of the overall financial, strategic, and structural elements of business, that emerge from a reading of the voluminous handwritten letterbooks that survive from his eight years as president of the Bell Company.

Under Forbes's command, Vail's influence on matters of company policy and strategy grew stronger. The two men had accumulated experience in major enterprises of national scope. Both had planned and coordinated large-scale, complex business oper-ations. Their shared perspective gave them a common approach to the National Bell Company's future: they stressed the need for clearly defined objectives and a systematic administration, and they tried to anticipate problems instead of merely reacting after difficul-ties had arisen. Hubbard's vision, like Alexander Graham Bell's, had been grand, and his instincts had been highly opportunistic and entrepreneurial. Under his part-time and often erratic leadership, the Bell telephone enterprise had been run in a fashion that was relatively loose, ad hoc and generally reactive to external events. The firm's new leaders subjected the company's affairs to closer, more systematic administration. With their greater financial re-sources, they were prepared to invest in management.

Administratively, the National Bell Company strengthened and standardized its relationships with its loose confederation of operating agencies. Forbes and Vail first requested clarification from the licensees of the written or verbal understandings they had relating to their contractual arrangements. Vail also began to streamline the flow of business correspondence. Traveling agents, in addition to Oscar Madden (who became superintendent of agencies), were hired and appointed to tour and inspect the licensees on a routine schedule "with a view to determining the best method of keeping accounts between the [Bell] Company and its agents, and the adoption of some uniform system for keeping the agents' accounts required to be fixed by the company . . . and to make settlements. . . ."[11]

The appointment of salaried traveling agents to cover the field of licensees provided a crucial managerial link between Bell and its operating agencies. The itinerant agents quickly became a source of intelligence on the state of competition and on a variety of local political and economic conditions that might have an impact on the establishment of exchanges. They also cross-fertilized the technology as they carried news of developments from one exchange to the next; improved the flow of information on technical problems and solutions between the Bell Company and the licensees; and even organized new Bell franchises. By 1880 the company was sufficiently educated on the common technical problems of the rapidly developing exchange business to be able, for the first time, to issue a printed pamphlet on the subject as a guide to new licensees.[12]

In general, the traveling agents systematized a practice that formerly had consisted of occasional and selective visits to the field by Watson and Vail. It was an important step toward achieving managerial control over the expanding business by making it easier for the Bell Company, sitting at the center of a greatly increased flow of information, to respond to specific problems with appropriate advice. Standard procedures could now be employed in dealing with common problems.

The National Bell Company was broadening its vision, looking ahead to the development of interexchange communication and to "connecting cities and towns" by telephone, a feat not yet technologically feasible beyond ten to fifteen miles.[13] After alleviating some of its financial and administrative problems, the company renewed its determination to expand its exchange operations. It would seek to gain control of as much of the market as possible in spite of the competition from Western Union. With an air of confidence, the company belittled the quality of Gold and Stock's telephone hardware and service, while publicly it asserted the superiority of the Bell patents. Vail shored up the morale of the agents by promising that if they were unable to tap new sources of funds, "our Company will immediately take steps to assist companies in all places. . . ."[14] Meanwhile, Forbes pursued negotiations with Western Union, ever hoping for a reasonable settlement that would leave the Bell interests in control of the telephone business.

Thus, from the spring of 1879 the Bell interests succeeded in loosening some of the financial and administrative constraints on their business. After the creation of the National Bell Company under wider ownership, they were able to marshal greater capital resources and to develop a more aggressive and efficient management, both of which were sources for a renewed confidence in the firm's ability to maintain a leading role in the telephone field. Other constraints on the business—the company's limited technological resources and productive capacity—now came into sharper focus. Successful growth of the company depended greatly on its ability to arrange for a speedy flow of well-built, technologically competitive telephones and apparatus. To this end, the new company abandoned the policy of depending on Charles Williams, Jr., for all of its production needs. A broader base in production and development was needed to match the firm's rapidly expanding networks of exchanges.

In the summer of 1879 Bell granted manufacturing licenses to four additional electrical manufacturers—in Chicago, Indianapolis, Cincinnati, and Baltimore—for the production of telephone apparatus. Williams alone was charged with the production of the telephone receiver and transmitter, an arrangement in keeping with the Bell Company's desire to control the distribution of the basic telephone instrument and with its faith in the quality of his work. The additional manufacturing capacity, it was hoped, would alleviate the pressure on Williams and enable Bell to supply the expanding market on a more timely basis.

The new manufacturing licenses were more limited in scope, less generous, and more carefully designed than the Charles Williams contract of 1878. They reflected the National Bell Company's continued determination to control the price, quality, and method of dissemination of all of its patented apparatus. The strict terms of the contracts also reflected a heightened awareness of the need to exercise close control over every aspect of the business as it spread geographically and grew technologically more complex. No longer would close personal ties of the sort that existed between Bell and Williams suffice. Relationships were contractual and hence carefully and fully defined. This, too, was consistent with the new type of managerial administration that Vail and Forbes were bringing to the telephone business.

· II ·

When Vail concluded on 17 February 1879 that "it is an absolute necessity that we arrange for some other factory to take part of the work," his decision was made amidst a rising crescendo of complaints from the field. "We should have a Western factory and probably it will be necessary to have one in New York," he wrote the next day, and on the twenty-first he informed Williams that he was now advising agents "to go temporarily outside and get supplies."[15] Given the number of complaints about orders, the wonder is that Bell waited as long as it did to expand the productive capacity on which it could call.

Once the decision had been made, Watson embarked on a tour of prospective manufacturers. By 28 February, he had concluded that Post and Company of Cincinnati, a manufacturer that functioned as a Bell operating agency for territories in Ohio, Indiana, Kentucky, and West Virginia, could supply the entire West. Watson felt that Davis and Watts, a Bell agency in Baltimore, along with Partrick and Carter of Philadelphia, could supply the South, and he proposed the establishment of a factory in New York.[16]

The directors of the newly organized National Bell Telephone Company decided the matter in principle on 21 March. George Bradley, the company's treasurer, reported that Vail was now authorized "to contract for such telephones and telephonic supplies as are needed to fill orders and also to maintain a stock on hand for ordinary contingencies."[17] With this broad mandate, Vail acted quickly to conclude agreements with manufacturers for patented telephone apparatus.

Given the uncertainties of Bell's competitive position in the spring of 1879, not every manufacturer was eager to enter into such a relationship. According to Thomas B. Doolittle, traveling agent for the Bell Company, "some manufacturers in Connecticut" were chary of "such a hazardous undertaking."[18] But enough were willing to take the risk, and by midsummer four firms, in addition to that of Charles Williams, Jr., were licensed to produce call bells, annunciator drops, and other designated equipment under Bell-owned patents.

The four new manufacturers were the Electric Merchandising Company of Chicago, Davis and Watts of Baltimore, Post and Company of Cincinnati, and the Indianapolis Telephone Company. These organizations had been selected from a number of potential firms already in the business of producing electrical supplies in general and telegraph and telephone supplies in particular. It was determined that these firms possessed machinery and employees appropriate to the task, had potential for expansion, and were financially sound.[19] One leading candidate, Partrick and Carter, which had been doing a steady business with Bell since 1877, was rejected, probably because of its continual complaints that it was short of funds.[20]

The first and riskiest arrangement was made with the Electric Merchandising Company of Chicago. Chicago was not only a critical market but also a major distribution point, pivotally located at the head of the nation's growing railway system and an important Great Lakes port. But northern Illinois was subject to very intense pressure from a rival Western Union agency, and Chicago was the home of the Western Electric Manufacturing Company, which made Western Union's Gold and Stock telephones. It was in fact a general agent for Western Electric, George Bliss, who in January 1879 had formed the Electric Merchandising Company. He had attracted the attention of Gardiner Hubbard that spring, and his budding enterprise had made a favorable impression. Hubbard wrote to Boston saying that Bliss carried a "fine stock of tools, one of the oldest telegraphic machinists in the country, and about $10,000 cash capital. . . ."[21]

The opportunistic Bliss had just rejected an offer from Western Union to become general superintendent of its subsidiary, the American Speaking Telephone Company. He had turned down a

healthy salary, gambling instead that he might strike a more lucrative alliance with the Bell interests. He hoped to promote more
than the manufacturing business. Offering to dispatch canvassers to
the principal cities of Illinois, he proposed the organization of a
"District Telephone Company" to be financed by local parties who
would give the Bell Company 40 percent of the capital stock. Bliss's
knowledge of the manufacturing arm of the competition and his
contention that he could lure two more experienced machinists away
from Western Electric made him a hot property. Vail convinced
William Forbes, the new president of the National Bell Company, of
the efficacy of doing business with Bliss, and a manufacturing
contract was concluded in June. A license for a telephone company
apparently did not become part of the package.[22]

Bell also needed capacity in the Southeast, where a company
agent had been canvassing Virginia, the Carolinas, Florida, and
Alabama since December 1878. Hubbard had already suggested a
possible manufacturing arrangement with a Baltimore firm, the
partnership of Davis and Watts, which produced electrical equipment. Davis and Watts invited Vail to inspect their manufacturing
facility in January. Little has come to light about the course of these
discussions, but by late spring Bell was under considerable pressure to
recruit this "southern" producer. The unexpected success in tapping
the southern markets made it essential to increase production, and
by the summer a licensing agreement had been concluded.[23]

The third new Bell licensee, Post and Company, enjoyed a
fine reputation as a prominent manufacturer of railway, telegraph,
and general electrical supplies. Watson had visited Post and solicited
sample instruments, which upon inspection proved satisfactory
enough to outweigh warnings from E. T. Gilliland (a former Post
electrician) that the company was "making outrageously horrible
work" in an "unsafe" facility.[24] Post in fact held valuable patent
claims for a magneto bell and could turn out fifty bells per day—with
projections in May for double that amount, enough "fully [to] supply
the demand in the West and Southwest."[25] Negotiations proceeded
to a successful conclusion, and although Post and Company subsequently caused some operational problems for the Bell Company,
Post turned out equipment whose quality was well regarded.

By licensing these three companies, Bell expected to expand not only its productive (and distributive) capacity but also its capacity for technological improvements. The most interesting, and perhaps most significant, step taken by the National Bell Company in this regard was its arrangement with a fourth concern, the Indianapolis Telephone Company. The Indianapolis Company was a recent Bell operating licensee, having been organized in February 1879. Its manager was former Post and Company electrician Ezra T. Gilliland, who was also an experienced telegrapher, an inventor, and a man of strong entrepreneurial instincts. Having secured the backing of powerful and wealthy local businessmen, Gilliland set about developing an exchange business in an attempt to stave off a budding Western Union effort in the Indiana capital. So successful was Gilliland that by June he had secured nearly two hundred subscribers and was "confident in his ability to hold Indianapolis against the W.U. . . ."[26] Much of Gilliland's success was attributed to his capacity for innovation.

Gilliland's main interests lay not in operations and service but in the development and production of hardware. At some point he had become a close friend of Thomas Watson's, a kindred spirit. In April he had told Watson that he was attempting to convert an old sewing machine factory into a "first class supply depot," and on the strength of Watson's recommendation, the Bell Company had taken a hard look.[27] Oscar Madden had visited in May and had been favorably impressed.

Gilliland's personal ties to Watson perhaps had an influence on Madden, but for whatever reason, Madden was convinced of the innovative potential of the midwestern manufactory. Especially interested in exchange technology, Gilliland worked assiduously on ways to outstrip Western Union inventions, about which (through his old telegraph contacts) he kept himself closely informed. His large, movable-peg switchboard, made in accordance with inexpensive methods of construction, almost immediately became the most popular among the Bell exchange agencies.[28] Even more immediately attractive to the Bell Company's management in the spring of 1879 was the Indianapolis Company's "considerable progress in working out cheap methods of manufacture." While Charles Williams, Jr., continued to manage his production much in the manner of a craft, Madden reported from Indianapolis that Gilliland had developed "facilities for turning out work which sets anything and everything which Williams has completely in the shades. . . . The machinery which Gilliland has enabled him by the use of unskilled and hence cheap labor to make all parts perfectly interchangeable."[29]

Citing as proof Gilliland's comparatively low prices for the production of Blake transmitters and call bells, Madden concluded that Gilliland was as much a potential threat as a potential opportunity for the Bell interests. His factory had already attracted the attention of Western Union, which was now making overtures. "I look upon G. as an exceedingly bright and ingenious mechanic, one who can aid or hurt us very materially," Madden wrote, "and for that reason should be kept in our service." Responding to the Bell Company's anxiety about the productive capacity of his plant, Gilliland assured Watson that his backers would "spare no expense" and that he would soon be geared up to manufacture one hundred sets per day. The plant was also being prepared for electroplating "on the grand scale" and the production of insulating wire.[30]

There is some evidence that the Indianapolis Telephone Company had not initially figured in the Bell Company's search for additional manufacturing capacity as directly as the Chicago, Cincinnati, and Baltimore firms. The Chicago arrangement, the first of the four to be consummated, on 11 June 1879, was a strategic necessity given the importance of the Chicago market. Arrangements with Davis and Watts and Post and Company, which were contracted in July, grew out of well-established associations based on those firms' roles as Bell operating licensees; their locations were also appropriate (as Watson had noted in February) to the geographic trends then evident in the expansion of the business. The Bell management's arrangement with Gilliland was more fortuitous. Once presented, however, the opportunity came to be seen as a good risk, full of promise. On 11 August, accordingly, the National Bell Company and the Indianapolis Telephone Company signed a manufacturing contract.

Thus, by midsummer 1879 National Bell had responded to the proliferating demand for telephone equipment in a geographically expanding market by departing from its reliance on a single manufacturing facility located in the extreme northeastern part of the country. Despite the security, good will, and de facto control implied by its exclusive relationship with Charles Williams, Jr., the pressures of growth and competition forced Bell to seek out new sources of supply. But having abandoned one principle, the Bell management stood firm on another. Charles Williams, Jr., was still to maintain his status as sole producer of the telephone receiver and transmitter under Thomas Watson's watchful eye. After due consideration of the alternative, Bell decided to license the new producers to manufacture patented telephone apparatus only.[31]

· III ·

The manufacturing licenses granted to the Electric Merchandising Company, Davis and Watts, Post and Company, and the Indianapolis Telephone Company were different in several respects from the one that Charles Williams, Jr., had signed in 1878. The new licenses were intended to preserve the Bell Company's control over most aspects of the patented equipment. This would be even more difficult now that the producers were scattered about the country and their several businesses were run by men who did not have strong personal ties (as Williams did) to Bell. No thought was given to the possibility that Bell might take an equity position in any of the producers' firms. Instead, they would seek to control production and distribution of apparatus through formal contractual ties. Strict, written agreements were devised to replace the relatively informal controls that had sufficed in the firm's early years. In this aspect of the business, as in others, individual differences were giving way to organizational needs as the business grew larger, spread farther about the country, and embraced a more complex technology.

The license contract negotiated with the Electric Merchandising Company on 11 June 1879 was the prototype for the three manufacturing licenses that followed. While the contracts specified the broad conditions under which patented equipment could be manufactured, the precise instruments to be produced and their prices were set by supplemental agreements signed on the same day. The licenses were handwritten and about ten pages in length (in

contrast to the three-page agreement with Charles Williams, Jr.).
The "standard" contract was divided into eleven parts:[32] The manufacturer was licensed to produce "instruments relating to telephony"
owned or controlled by the National Bell Telephone Company. The
specific equipment to be produced was to be designated in writing
"from time to time." Because Bell hoped to begin earning income on
the sale of auxiliary apparatus (income that currently accrued to
Charles Williams, Jr.), the manufacturers were to pay an annual
license fee of not more than 20 percent of the amount for which each
instrument was sold to the Bell Company's designated agents, except
in cases where the Bell Company itself had to pay an inventor's
royalty or license fee. In such cases the royalty or fee was to be borne
by the manufacturers, regardless of its proportion to the instrument's
sale price. Such royalties or fees were to be charged equally to all
manufacturers. For the present, however, the supplemental agreements waived any royalties or license fees "until further notice,"
most likely as an incentive for the new producers to achieve a high
output as quickly as possible.

As executed, the new manufacturing contracts reflected the
Bell Company's concern to expand production without risking
control over its patent position. The basic telephone instrument, the
company's primary patent base and principal source of revenue,
remained sacrosanct. Even though the contracts seemingly extended
the purview of the manufacturing licensee to all telephone equipment, "apparatus or instruments" alike, the supplemental agreements, which specified the equipment to be made, confined production to apparatus only. In the supplemental agreements the four new
manufacturers were authorized to make magneto call bells, electro,
or "hook," district bells, and annunciator drops, on which the Bell
Company held patents. Davis and Watts and the Indianapolis Company were authorized to produce office bells, and Post and Company
was licensed to make some Bell-patented exchange apparatus.

Technological concerns of the Bell Company with regard to
quality, innovation, and patents also loomed large in the contracts.
To meet adequate quality production standards, the manufacturers
agreed to submit their products to Bell Company inspection. In the
event of "improvements" on apparatus made by the "agents, employees, or servants" of the manufacturers, they were to be conveyed

exclusively to Bell for a "fair price" (which if not agreed upon would be submitted to the arbitration of three referees). To further protect its patent position, the Bell Company specified that all equipment produced under the license contracts be furnished only to authorized Bell agents and under restrictions set by the company "from time to time." The manufacturers were required to number all hardware in series and to label each piece of equipment with the words "Made for the National Bell Telephone Company," along with the name and address of the producer. Strict accounting procedures were put in force, requiring manufacturers to furnish weekly statements of the amount of finished equipment produced, the number of orders (and by whom) for each category of equipment, and the number of orders filled (and to whom), with dates of shipping.

Finally, the contract attempted to secure for the Bell Company absolute control over the pricing of patented apparatus. Control over prices reflected a complex of concerns. First were the Bell Company's specific desires to ensure minimum production standards and to allow for discrimination in prices that would favor licensees in which the company had taken a financial interest. Second was the general propensity of contemporary businessmen to try to mitigate price competition in the deflationary economy of the 1870s. The prices—which were to cover the cost of manufacture plus license, or royalty, fees (if any) plus "a fair profit"—were to be uniform among producers and subject to change no more often than every six months. To this end, the strict accounting requirements of the contracts asked for annual statements under oath of "full and true returns." Initial prices for the major items were established as follows: magneto call bells, $8.00; electro, or "hook," bells, $3.25; office bells, $2.50; annunciator drops, $1.50. One exception to this schedule of prices is found in the agreement with Post and Company, which agreed to ship lots of five hundred or more instruments of each kind directly to the Bell Company at a discount.

More than any other requirement of the 1879 manufacturing contracts, the constraints on pricing would prove to be the most problematic and the least amenable to enforcement. For the time being, however, Bell hoped that this and other provisions of the contracts would be adhered to under threat of revocation of the license on simple written notice. Otherwise, the life of the contracts was five years from the date of execution, cancelable by either party on six months' notice.

The differences between these contracts and the one negotiated a year earlier with Charles Williams, Jr., are significant.[33] Under his contract, Williams alone retained the right to produce the telephone (receiver and transmitter). Williams had also been granted certain other commitments that were not given in explicit terms to the other manufacturers. He alone was entitled to limited indemnification for any losses sustained because of innovation; he alone was allowed to draw money on consignment for the call bells that he made; and he alone was promised in writing that the Bell Company would promote his products among the licensed agents.

On the other hand, Williams seems now to have been expected to conform to the provisions of the 1879 contracts requiring the filing of reports, the labeling of instruments, the payment of royalties or license fees, and the furnishing of equipment to Bell licensees only. Operational instructions from Vail to the manufacturers from the summer of 1879 onward applied equally to Williams, although no one felt the need to renegotiate the August 1878 license.[34] The provision in Williams's contract allowing the Bell Company to terminate the license on ninety days' notice may have had something to do with his acceptance of this new situation, but it is more likely that the maintenance of the status quo was a result of continuing good relations between the two firms. Despite the new contracts, the manufacturer and patentee were still bound by an association that stemmed from the very genesis of the telephone, by their physical proximity, and by their mutual confidence in the quality of the Bell telephone.

· IV ·

The Bell Company seemed certain that its array of manufacturing licensees would enable it to meet the rising demand for telephones and turn back Western Union's challenge. Spread around the country as they were, the new manufacturers were expected to alleviate some of the delays in shipping. The company hoped that the contracts would enable it to guarantee production of high-quality equipment and at the same time secure to Bell expected "improvements" in equipment design. The provision fixing prices equally for all manufacturers was intended to prevent price competition, leaving quality of product as the competitive variable. While the Bell Company would soon extract an income by exercising its option to charge licensing fees, the manufacturers alone reaped the profits from the sale of call bells, annunciator drops, switching devices, and any other apparatus that they produced for use with the telephone.

As for quantity of production—the problem that had triggered the search for additional capacity in the first place—the Bell Company's expectations were still influenced by its faith in Charles Williams, Jr. By withholding telephones from the supplementary agreements to the new license contracts, the Bell Company continued to limit the quantity of telephone service to the capacity, albeit enhanced, of the Boston shop. Williams would now have every opportunity to improve telephone production, freed as he was from the obligation to make most of the other apparatus. (Williams continued to make patented telephone apparatus, but mainly, it seems, for the New England–New York market.) This arrangement promised to preserve Bell's close control of its most important patented product, while it enabled Bell at last to meet all the demands for equipment coming in from the agents. And although there was no formal guarantee that the licensed manufacturers would increase output on demand, there was at least the expectation that the new producers would agree to "turn out instruments rapidly."[35]

Four other key assumptions surrounding the new manufacturing arrangements were not made explicit in the license contracts but can be inferred from them and from the contemporary correspondence between the Bell Company and its manufacturing licensees. These assumptions were: (1) that the manufacturers would compete to some degree for the business of the licensees; (2) that quality, not price, would be the chief competitive weapon; (3) that regional producers would tend to serve regional markets; and (4) that there was no need for rigid standardization in apparatus design.

The most dubious aspect of these assumptions was that markets would tend to divide along geographical lines under competitive conditions. At the same time, the license contracts of 1879, unlike the Williams contract, carried no commitment from the Bell Company to promote a manufacturer's business. As Vail put it to Post and Company in April, "What we proposed to do in making arrangements with you for manufacturing was to give you licenses to manufacture under our patents, provided you would sell telephonic apparatus only to our agents and keep the work up to a certain standard. You of course taking your chances of obtaining work from our agents. All that we proposed to do was to recommend you to the same."[36] Vail went on to explain that sheer demand would provide "enough in the telephonic business of the future for all parties . . .," assuming that the manufacturers confined themselves to regional markets. This certainly had been Watson's thinking in February, and Post and Company suggested that "the best interests of all concerned would be better served by limiting the Manufacture . . . to say 3 and at most 4 Concerns with a proper division of Territory."[37] Still, the Bell Company left open the opportunity for its leasing and operating agencies to exercise some choice in the procurement of telephone apparatus. Market boundaries were not rigidly and explicitly defined, and soon this would lead to problems.

A major factor behind the company's refusal to grant explicit rights to specific regional manufacturing territories was the need to encourage variety in telephone apparatus at this early stage in the industry's development. The tolerable range for variations in the auxiliary apparatus was sufficiently wide to allow for independent tinkering and distinctions in styling. This was especially desirable for the Bell Company, which had as yet no formal approach to research and development. Beyond the insistence that equipment be produced up to "a certain standard," Bell had no rigid specifications to hand out to the manufacturers. Thus we find Vail on one occasion writing that he "prefer[red] to have each manufacturer follow his own design" in making magneto bells and on another occasion asking Gilliland to attempt to "combine in one bell, the best features of all."[38] Competition along these lines would stimulate innovation. Granting exclusive territories would have made it difficult for some manufacturers to satisfy some of their customers. Bell was willing at this point to leave that sort of relationship to be worked out by the "invisible hand" of the market.

Competition had not only virtues but also limitations. Inter-firm rivalry among the manufacturers was bound by the Bell Company's insistence on setting prices for telephone apparatus. The reasons for this were complex. As we shall see, the company wanted to ensure that its favored operating licensees, those in which it held equity, would receive apparatus at a discount, which was possible only if overall prices were held to a reasonable minimum. More important, Bell simply did not want its manufacturers to engage in sharp, competitive price warfare. Quite aside from the adverse affects price warfare could have on the financial health of the manufacturers Bell had worked so hard to recruit, such competition threatened to disrupt the smooth functioning of a tripartite relationship that required careful coordination of limited resources. Price competition, furthermore, would be detrimental to the quality of the product. Vail put it simply: "Our idea is to have the prices uniform and if there is any difference at all, it must be in the superiority of the workmanship."[39] Price cutting was the most serious offense a manufacturer could commit against the harmonious working of the system. An attempt was made, therefore, to set prices high enough to guarantee a "fair profit" for the manufacturer yet low enough to assure the operating agents a good return when they provided the telephone equipment to the subscriber. We shall see, later, how Bell's expectations for regulating its manufacturing relationships, in this and other regards, were undermined by the uncontrollable realities of competition.

· V ·

The National Bell Company's new manufacturing arrangement was a response to the growth and geographic expansion of telephone demand, but it also signaled an underlying shift in the orientation of the Bell entrepreneurs. The extension of production licenses to several manufacturers was, after all, occasioned by the needs of its operating licensees. Bell's growing concern for the health of the agencies, especially those in which it held stock, was reflected in changes in the firm's approach to pricing telephone equipment.

The very existence of several manufacturing licensees invited greater attention to producers' prices. From 1877 to mid-1879 Bell had relied almost entirely on Charles Williams, Jr., for an assessment of production costs. The markup on Williams's costs was limited by conventional standards as to what constituted a "fair profit" and by his close working relationship with the Bell management. The recruitment of new manufacturers, however, provided the Bell Company with a broader basis for assessing production costs. Now the burden of proof could be shifted to potential producers, who had to demonstrate their ability to make apparatus at reasonable cost. When Madden visited Gilliland in late May 1879, for example, the latter's proposed low prices were a revelation. Some form of cost control seemed to be a necessary innovation.

Although the data are fragmentary, it is possible to get at least an impressionistic sense of how telephone equipment prices were related to costs at this early date. We must bear in mind, however, that cost accounting in nineteenth-century manufacturing plants was unsophisticated by modern standards. The pricing of finished goods was usually related to "prime costs," that is, to the aggregation of the direct costs of labor and of materials purchased and consumed. General overhead expenses were then added to determine the price. The category "general expenses" was, however, vague and variable, often determined by rule-of-thumb estimates of the cost of replacing the plant, along with a subjective judgment about the level of a fair profit. These were added to the totals for more readily identifiable overhead expenses, making prices very much an ad hoc phenomenon.[40] The range within which a wholesale price related to costs could be very wide, especially in the absence of competition.

For example, until March 1878 Charles Williams, Jr., stated that he added 5 percent to the cost of material and then added "a certain amount in advance of what I pay the men per day" (varying from sixty-five cents to one dollar above each worker's daily wage) in setting the price of his telephone equipment. This presumably amounted to less than the 33–50 percent that other manufacturers allegedly added to their prime costs, since Williams thought that his return was lower than the "industry" standard.[41] The Bell Company directors felt that a fair return could be set equal to prime costs plus

15 percent general expenses, although in practice the returns may have been higher on certain kinds of apparatus. By April 1878 Gardiner Hubbard noted that the cost to Williams for producing a call bell was eight dollars, making it possible to sell at ten dollars, or a 25 percent markup.[42]

This led to some complaints that Williams's prices were excessive. In June, one agent argued that the Bell company was "paying 25 to 50% more than [it] should for instruments." By November new traveling agent Oscar Madden was urging that lower prices be quoted to agents, saying that the company's "premises as to cost and longevity of . . . instruments" must be "materially modified." He suggested, nevertheless, that a 25 percent markup was a fair return for the manufacturer. As it happened, the Bell Company closed the year paying Williams $3.10 for each telephone that he produced at a cost of $2.70. This was a markup of 15 percent, precisely the return that the Bell directors had deemed fair earlier in the year. It was also 3.5 percent less than what Williams had requested and 4 percent more than he had said that he could afford.[43] Thus, when the Bell Company began to canvass additional manufacturers in the spring of 1879, there was already some downward pressure on telephone prices. After Post and Company placed a bid for telephones at $2.60 each, Watson warned that "brother Williams will have to come down."[44]

Learning helped. As Williams gained experience, he was able to cut the production costs for telephone equipment, although charges for certain items rose at times, owing to elaborations in design or improvements in quality. Thus, while the cost of producing the telephone increased at the end of 1878 with the advent of the Blake transmitter, the cost of call bells declined significantly from 1877 to 1879. Hubbard noted that between January and April 1878, the cost of the magneto bell dropped from twelve dollars to eight.[45]

Gradually, considerations of production costs became more important to Bell. At first they had been ancillary to the drive to increase output without sacrificing the company's control over the distribution of telephones. Gilliland's claim to be able to produce a Blake transmitter cheaply did not persuade Bell management to depart from its sole reliance on Williams to make the telephone.[46] But the multiple manufacturing arrangement did provide the Bell

Company with a new and useful opportunity to assess costs of auxiliary apparatus by inviting comparative estimates. This was done not only at the beginning of the contract period but at each price revision between 1879 and 1881. Although Bell derived no direct monetary benefit from the sale of apparatus to the licensees, it had much to gain by keeping costs (and thus prices) as low as possible. In this way Bell could provide the manufacturers with the "fair profit" that they were assured, while providing affordable equipment that would contribute to the expansion of the market for Bell's telephones.

As the Bell Company became more heavily involved in the business of its operating licensees, more direct financial benefits began to result from cost cutting in production. After 1880 the firm recognized the desirability of lowering telephone equipment prices to exchanges in which it held an equity interest. It did this by offering select exchanges discounts on the basic telephone instrument of 30–50 percent.[47] On apparatus, discounts were applied in January 1880 to the exchanges in Boston, New York, and Chicago— exchanges now wholly or partly owned by the Bell Company. The prices were set as follows:[48]

	Magneto Bell, Automatic Switch	Magneto Bell, Secrecy Switch
General price	$8.50	$8.75
Price to select exchanges	7.50	7.75

Toward the end of 1880, when improvements in the apparatus and the introduction of additional product lines prompted another revision of prices, the manufacturers were asked for a quotation on what each considered to be a fair price for call bells and what discounts they would allow for nine select exchanges. The prices were set so that four varieties of call bells ranged from $8.25 to $11.63, from which one-third was subtracted for nine exchanges.[49] This price discrimination clearly was intended to enhance the profitability of those exchanges in which Bell had acquired an ownership interest, and yet the absence of complaint from the manufacturers seems to indicate their satisfaction with the overall level of prices.[50] The absence of objections from the unfavored operating agents to the subsidization of Bell-owned exchanges can be attributed to the fact that everyone concerned was enjoying a healthy business.[51]

Theodore Vail believed that the prices of telephone equipment were fair to the manufacturers and yet low enough to enable the agencies to make good profits. In July 1879 he explained to W. O. Rockwood, Gilliland's patron, that "it is to our interest that we should put the manufactured apparatus to the agents as low as is consistent with good workmanship, and proper interest on the part of manufacturers." The manufacturer would never be charged a royalty, he explained; instead, it would be passed on to the customer. This is precisely what happened when the Bell Company ordered that investors' royalties be charged to the cost of manufacturing in November 1879; prices were raised accordingly.[52]

Summarizing the manufacturing arrangement in March 1880, a Bell Company memorandum boasted:

> [W]e have endeavored with success to keep the cost of all this auxiliary apparatus and supplies at as low a price as possible in order to facilitate the introduction of telephones. Our efforts in this direction have proven very successful, and the result has certainly shown the wisdom of the policy. The cost of the electrical apparatus used by our agents is barely 50 per cent of that of similar apparatus in use for telegraphic purposes, when the telephonic business began about 3 years ago.[53]

A year later, Vail reiterated the point that the company's policy was "to make all the auxiliaries to the telephone, as cheap as possible."[54]

Thus, two years after the Bell Company had decided "to go temporarily outside [Williams] and get supplies," the firm's approach to costs and prices had changed in significant ways. In the first place, the company became more sensitive to the ongoing problem of adjusting prices to considerations of demand and quality, both of which were related directly to the costs of production. The existence of several manufacturers allowed for a more objective, comparative basis for ascertaining the cost of apparatus. Simultaneously, the company was becoming more sensitive to the needs of operating agents as purchasers than to the needs of manufacturers as sellers. This reflected in part a simple desire to expand the market; it was also a result of the competitive and technological pressures that had forced the Bell Company to establish links with some of its more important urban franchises through new ties of ownership. In these and other ways, attempts to control various aspects of manufacturing, including the prices of hardware, were tightly interwoven with an incipient process of integration forward into the operations side of the business.

· VI ·

Had the Bell interests not been faced with serious competition in the spring of 1879, they might not have felt as pressured as they did to increase the production of telephone equipment through the licensing of multiple manufacturers. Somewhat ironically, just as the new manufacturing arrangement was being implemented, the industry's competitive setting was altered drastically. This happened when Western Union and National Bell arrived at an agreement to end the sharp competitive struggle that had contributed so much to the shaping of early Bell strategy. Just why the forty-million-dollar telegraph giant (which had, according to Theodore Vail, five hundred thousand dollars to spare for the telephone business) dropped its competing patent claims to the telephone is one of the more curious and fascinating questions in the history of American business.

Writings on the subject offer two fairly simple lines of interpretation.[55] The first is that Western Union abandoned the field once it became convinced of the legal weakness of its position. This argument is based solely on friendly testimony given in 1881 by George Gifford, Western Union's chief counsel, in a subsequent Bell patent suit. Gifford maintained that by June 1879, after testimony in the patent suit that Bell had brought against his clients nine months earlier had closed, he had become

> convinced that Bell was the first inventor of the telephone and that the defendant . . . had infringed said Bell's patent by the use of telephones in which carbon transmitters and microphones were elements, and that none of the defences which had been set up could prevail against him; and I advised the Western Union Company to that effect, and that the best policy for them was to make some settlement with the complainants.[56]

On the strength of this definitive legal advice, the story goes, Western Union agreed to settle out of court the infringement suit that Bell had brought against it in September 1878.

A second, complementary interpretation holds that the Bell Company was the beneficiary of an unusual historical accident.[57] In 1879, while telephone competition with Bell was in full swing, Western Union was being threatened on its flanks by the aggressive financier Jay Gould. Gould, one of the notorious robber barons of the day, was mounting a full-scale assault aimed at seizing financial control of the company from William H. Vanderbilt, eldest son of Cornelius, Gould's bitter enemy in the legendary struggle for control of the Erie Railroad. In May, Gould formed the American Union Telegraph Company, and by the end of the month he was making serious incursions into Western Union's telegraph business. Gould, moreover, was actively purchasing interests in Bell telephone exchanges in Connecticut and was making overtures to Vail for a combination of interests. The overtures apparently had some effect. Vail responded confidently to a licensee's complaint about Western Union's refusal to receive messages from his exchange: "You can say to your customers that before Fall there will be an opposition telegraph reaching to all the prominent points in the country, in complete working order and it strikes me with that opposition in the Fall the W. U. will only be too glad to be allowed to receive messages from your exchange."[58] By June Western Union had approached the Bell Company with a proposal for consolidation in order to prevent any alliance between the Bell and Gould interests.[59] This then set in train the negotiations that resulted in the signed agreement of 10 November 1879. In accordance with that understanding, Western Union divested itself of its telephone patents, giving the National Bell Company a virtual monopoly of the nation's telephone business.

The Jay Gould story has been offered by scholars working for the Federal Communications Commission (FCC) as the key to understanding the otherwise "surprising capitulation of the powerful Western Union to the diminutive Bell Company."[60] But, in fact, before Gould's entry onto the scene, the two firms were preparing to reopen negotiations for an end to their competitive struggle via a consolidation of interests. The owners and managers of Western Union did not like competition any more than their counterparts at Bell did. The underlying reality was Western Union's desire to preserve its monopoly over telegraphy (that is, the written transmission of intelligence) and its coordinate desire to ensure that the telephone was restricted to uses that would enhance, not interfere or compete with, the lucrative business correspondence commonly borne by the telegraph. The advice of counsel and Gould's machinations may help explain why Western Union stopped insisting on a combined telephone company in which it should hold a majority stock interest. But neither Gifford nor Gould was an essential actor in a scenario whose final act was the 10 November agreement, which was designed very much to Western Union's advantage.

The agreement, which brought to an end the Bell–Western Union infringement litigation, commonly known as the Dowd suit, was modeled after a smaller-scale licensing arrangement executed that summer between a Bell licensee, James Ormes, and the Gold and Stock Company's agent for the southern United States, D. H. Louderback. Ormes and Louderback had presented Forbes with a proposal to establish district telephone exchanges under terms that would eliminate competition in seven southern states. Under a tripartite agreement executed on 26 August, Ormes received an exclusive ten-year license from National Bell to establish district exchanges in his territory. For every telephone he rented, he was to pay Gold and Stock a royalty of one dollar in exchange for the latter's promise to refrain from competing. In cases where Ormes and Louderback might open joint exchanges, they agreed to use only Bell telephones. To satisfy Western Union's desire to protect its telegraph business from competition, the Bell Company agreed that all telegraph messages collected by its southern exchanges would be turned over to Western Union exclusively. In addition, Bell exchanges were to refrain from connecting with any other telegraph company.[61]

The Ormes-Louderback alliance and Western Union's ongoing attempts to solicit similar agreements with other Bell agencies must surely have impressed the Bell Company with the need to pursue a general settlement. Vail expressed concern over the "detached arrangement" struck by Ormes and Louderback, although he recognized its value as a basis for future agreements.[62] The Bell Company could hardly afford to have all of its operating agencies striking different agreements, especially at a time when Bell was attempting to place all of its licensing arrangements on a more consistent, uniform basis. The best solution was to follow the Ormes-Louderback model and reach a settlement between the two major competitors.

That was precisely what was done in the 10 November agreement between Bell and Western Union. In retrospect, we know that Western Union made a strategic blunder, but from a contemporary perspective, the telegraph giant fared quite well. In return for its assignment of all of its telephone rights (and patents) to National Bell, Western Union received concessions that made secure its dominance of the telegraph market. Telephone exchanges were generally limited to a radius of fifteen miles from a central office, and any telephonic connection between them was limited to the purpose of "personal conversation" only. Such connections were "not to be used for the transmission of general business messages, market quotations, or news for sale or publication in competition with the business of the Western Union. . . ." The Bell Company, moreover, agreed to transfer all telegraph messages to Western Union. The crowning concession was Bell's commitment to pay Western Union a 20 percent royalty on the rental income from every telephone that it leased in the United States for seventeen years. The agreement thus allowed Western Union to reap a handsome direct profit from the telephone, eliminate a source of competition in its major business markets, and gain an ally in its fight against other telegraph concerns.[63]

In structural-functional terms, the agreement could be seen as Western Union's way of assigning to the Bell Company the problems of managing the local loop, voice-grade communications systems attached to Western Union's nationwide telegraph network. With the market separated into local and long-distance parts, Western Union stood to benefit in the following way. As a captive feeder service, Bell's business would enhance Western Union's telegraph traffic and revenues. Western Union, moreover, was freed from financing and managing the relatively capital-intensive and technically complex telephone exchange operations. With the Bell Company taking on the burden of providing feeder services, Western Union could expect to increase the overall value of its long-distance service in a way roughly analogous to contemporary arrangements struck by major railroads and smaller lines—arrangements that left to the latter the worries and costs of managing the less profitable, short-haul business.

Despite the apparent advantages accruing to its former rival, Bell could afford to celebrate this climax to a year that had begun with gloomy prospects. The settlement gave the firm an important measure of security and greatly enhanced the value of its holdings. Almost immediately Bell stock doubled in value.[64] With greater capital resources and a diminished sense of uncertainty, the company was in a good position to extend and consolidate its telephone business. The agreement transferred to Bell and its licensees valuable assets in plant and equipment and all of Western Union's rights to eighty-four patents on telephones and apparatus. Western Union agreed to make its "lines, poles and structures" available to Bell and to pay it a 15 percent commission for the transfer of any messages to Western Union. Western Union telephone exchanges except those within the Ormes territory and in New York City were to be transferred to Bell licensees at fair market valuation.[65]

Thus, while the telephone business was in some important respects circumscribed in its ability to compete with telegraphy, it became virtually secure in its own domain. While Bell would continue to fight numerous patent challenges over the next two decades, there was no one left to mount a serious challenge to its rapidly growing and by now extensive series of telephone exchanges. No doubt the Bell directors were as satisfied as Norvin Green, president of Western Union, to be rid of "a bitter and wasteful competition."[66]

· VII ·

In the wake of the settlement, Bell could turn its attention to promoting further growth in the industry, further improving the technology of telephony, and further administrative consolidation of its licensees. On the production side, the 1879 manufacturing arrangement succeeded in the first two instances but taxed the administrative capabilities of the Bell Company as it tried to reconcile its policies of closely regulated pricing and distribution with pressures on the producers to compete for licensee business.

The firm's hope that its new multiple-licensing arrangement would alleviate its short-term supply problem was fulfilled. Complaints from licensees about the quantity, quality, or timeliness of delivery of supplies diminished. After November the absorption by the Bell licensees of Western Union exchanges and agencies was accompanied by an influx of Gold and Stock equipment, which was phased out and replaced with apparent ease over the next few years. By early 1880 the Bell Company was able to reject with confidence the application of a well-reputed manufacturer in New York to produce call bells, annunciators, and switching devices; as Bell explained, its productive capacity was sufficient. In January 1880 Oscar Madden reported that the company had a large stock of telephones and that orders were being filled on the day of receipt; toward the end of the year he proclaimed, "We have now fully as many licensed manufacturers as the business will justify."[67]

The company's increasing telephone inventory was not a result of slowing demand. The company may have anticipated a "large decrease in orders" after the Bell–Western Union settlement, but this, Vail noted, had not occurred. There was "evidence," he reported, "that the past demand is healthy, and was not created by competition and rivalry." Although Vail was far too sanguine in his projections for 100,000 new telephones for the year beginning in March 1880, demand nevertheless reached unprecedented levels. The manufacturing licensees were now able to meet the orders for telephone apparatus.[68] Expansion of the Bell telephone business through the second half of 1879 had involved placing only about 16,000 new Bell telephones into the hands of licensees, some 500 fewer than during the first half of the year. This may have been owing, in part, to the company's promise to Western Union not to open new exchanges after June, while negotiations for a settlement went forward (although this promise was frequently honored in the breach). But after 1880, demand swelled. Bell telephones in use more than doubled, from about 55,000 to 111,000 between 1 February 1880 and 20 February 1881. Charles Williams—as well as the new licensees—was able to keep pace with this rapidly growing market.[69]

Factors contributing to Williams's increasing output are hard to discern from the extant sources (which yield no information on the internal workings of his shop, its financial health, or its size in this period), but two things are certain: Williams's shop was growing; and he was able to increase his telephone production for the United States from approximately 670 per week in 1879 to approximately 1,000 per week in 1880.[70] He was no doubt helped by being relieved from the burdens of making, selling, and distributing a large volume of telephone apparatus, along with the associated administrative and bookkeeping tasks involved. Telephones were a relatively simple matter. Once produced, they involved only a simple transfer to a guaranteed buyer, the Bell Company, which in turn handled all the problems of distribution, including inspection.

The 1879 manufacturing arrangement also met the company's expectation that the several carefully chosen producers could make apparatus of good quality and help Bell to promote more effectively the technical improvement of its products. Staying abreast of the industry's rapidly changing technology was a growing concern. On the one hand, the National Bell Company continued to rely heavily on its own personnel for technical improvements. In May, Watson was appointed general inspector, with a hefty increase in salary, to preside over a growing staff of technical employees. By the end of the year this group included two electrical engineers.[71] On the other hand, the establishment of business relations with several electrical shops gave Bell access to a much broader range of individuals and resources involved in telephone technology. It was a reasoned response to the burgeoning interest in electrical communication in general and in telephone technology in particular. The company now benefited from a broader spectrum of ideas developed by experienced electricians and machinists, whose personal quest for patents would redound to the benefit of the Bell company.

In several cases the multiple-licensing arrangement stimulated innovations in telephony almost immediately. Gilliland's factory (in Indianapolis) not only employed novel approaches to machine production methods but was strong in small switchboard development; and Gilliland's suggestions for improvements on the basic telephone instrument (which he was forbidden to produce commercially) were surely conveyed with haste to Charles Williams,

Jr. Williams, in turn, continued to revise the telephone. Post and Company, which among the licensees earned the strongest reputation for the quality of its call bells, was also working at its own initiative on problems of resistance on long wires. As it soon became clear, some of the producers were more aggressive and innovative than others. As a result of these differences, the appearance and quality of equipment varied somewhat from shop to shop, leading many agents to prefer one factory's products over another. But while the Bell Company was always concerned that apparatus meet sound quality standards, it also found competitive variations in equipment design useful for the development of the technology.[72]

This success in devising technical improvements and in meeting demand was offset to some degree by the problems that arose in the administration of the 1879 licensing arrangement. The most nettlesome difficulties in managing relations with a far-flung group of independent manufacturers stemmed from Bell's need to enforce the contractual obligations related to pricing, filing reports, and restrictions on the sale of patented equipment. Less explicit in the day-to-day operational correspondence but obvious nonetheless was the problem of territorial overlap, a situation that suggests inefficiencies resulting from cross-hauling delivery patterns. The root cause of the troubles was the conflict inherent in the company's desire to have the apparatus manufacturers compete for the business of the operating licensees but to do so under conditions of strict price control. For their part, the manufacturers wanted a guarantee of patronage from the Bell operating agencies. Vail insisted that licensed manufacturers sell only to licensed agents, but he would not establish a reciprocal agreement. He promised only to recommend to agents that they buy equipment from the nearest producer. To Davis and Watts, for example, he explained, "If you close a contract with us, we shall recommend you to all our agents in the southeastern sec[tion] of this country in order that they may obtain their supplies from you. If you make an instrument in all respects equal to that made by other manufacturers you will obtain the bulk, if not all of their trade, and any additional trade that may come to you from other quarters."[73]

Vail soon discovered, to his chagrin, that not all the manufacturers were content to stake their business on the quality of their equipment. Price competition between the Indianapolis Telephone Company and Post and Company—for business on the fringes of their adjacent territories—surfaced almost from the moment the ink was dry on their contracts. On 9 August 1879 Post and Company accused the Indianapolis Telephone Company (renamed Gilliland in June 1880) of underselling both magneto and electro bells, thus "ruining the business by cheapening prices and product." This was a charge that the Indianapolis Company secretary, C. B. Rockwood, denied but failed to lay to rest. Bliss subsequently reported that his Chicago business was being hurt, since the local Bell agent was receiving annunciator drops and magneto bells from Indianapolis at an illegal discount. He also reported that the Milwaukee agent was on the illicit Indianapolis pipeline. Both the Post and Indianapolis companies, meanwhile, were exchanging accusations which both vehemently denied. Soon, however, the charges were confirmed, and the guilty parties were forced to settle their differences, for the moment at least, by agreeing to sell at the contracted prices.[74]

The problem did not vanish. By March 1880 Madden was writing to Post and Company to say, "We regret exceedingly your indisposition to maintain with fidelity the rates which all our manufacturers are obligated to adhere to." A few weeks later, Madden, as the newly appointed assistant general manager of the Bell Company, threatened the Indianapolis Company with revocation of its manufacturing license because of further transgressions of the pricing policy. The General Manager's Letterbooks for the next two years are filled with the written record produced by these struggles. The letters were to no avail. In the summer of 1882 Post and Company was discovered to be offering equal exchanges of new bells for old, boxing equipment without charge, and offering unauthorized price discounts.[75]

Evidence of the persistence of violations was clear enough. Vail grew weary of "continually calling . . . attention to these evasions of the conditions of [the] contract," but he seems never to have found an effective means for bringing the transgressions to a halt. Nor could Madden do much in November 1880, when Gilliland prematurely announced a systemwide price reduction. As the manufacturers in tandem lowered their charges, Madden noted helplessly that "the damage is . . . done."[76]

The obvious disadvantage of the pricing policy as far as the Bell Company was concerned was that it was almost impossible to enforce. Bell's only recourse was the drastic measure of revoking a license. From the vantage point of Charles N. Fay, the Bell agent for the Chicago exchange, the policy had other ill effects. Admitting that "we can buy bells cheaper from several parties out here than Mr. Williams prices," he confided that "all these manufacturers are ready to cut prices on each other right and left." Why should he not, therefore, "buy the best bells I can find at the cheapest price"? As far as Fay was concerned, the Gilliland bell was not only less expensive but "the best bell for power and beauty." In response to what must have been a suggestion that he do more business with Williams, Fay charged that Williams's bells were more expensive, of lower quality, and more costly in terms of freight and repairs. Then he pointed to the central flaw in the system. If the Bell Company wanted each manufacturer to enjoy an equal opportunity to sell its goods to any of the agencies, then "you ought to allow free competition in these matters. . . . I do not see how you can possibly allow manufacturers to compete in [the] same territory, and set artificially an arbitrary price irrespective of merits or cost of transportation."[77]

Indeed, the Bell Company was not consistent on this point. Its original conception of manufacturers serving regional markets inevitably conflicted with the principle that agencies should have a free choice in purchasing equipment. Price competition was one side effect; cross-hauling was another. Far-flung transactions between manufacturers and agencies seem not to have been uncommon. Post and Company, for example, sold call bells in Boston, New York, and Chicago; and as we have seen, Gilliland did business in Chicago and Milwaukee. George Bliss may have been disturbed that he could not sell to the local Chicago exchange, but he apparently was able to undercut Davis and Watts in Richmond, Virginia.[78] Oddly enough, no one in the Bell Company raised a question about the inefficiencies or costs inherent in cross-hauling, but clearly this had a deleterious effect on the morale of the manufacturers. Understandably, they complained when they thought their territories were being unfairly impinged upon.[79]

Another problem with the license contract resulted from the temptation for manufacturers to sell patented equipment to unauthorized parties. Manufacturers were reminded of their obligations under provision 3 of their licenses in March 1880 and again in August. In September Bliss was specifically admonished, but by the end of the year, Vail had received information that Bliss's Electric Merchandising Company was furnishing "about all the magneto bells used in connection with infringing telephones in Chicago. . . ."[80] Warnings to the manufacturers were issued many times through mid-1881, but for the most part these warnings seem to have been ignored. Moreover, the manufacturers were attempting to ship equipment abroad, often by surreptitious means. Even Williams had to be warned to stop selling telephones for export.[81] All of these practices were hard to detect and even harder to prevent. The Bell Company now found it difficult to monitor the production of equipment because of tardy reporting. Davis and Watts especially displayed "a sad want of promptitude" in this regard.[82]

· VIII ·

Years later, Vail and Charles Scribner, chief engineer of Western Electric, offered retrospective views on the Bell Company's licensing arrangement with these several manufacturers. Their views are important because they have formed the basis for the conventional understanding of the problems arising from the arrangement. According to Scribner, the essential problem was poor quality of equipment. "The Charles Williams, Jr. shop," he wrote to E. J. Hall in 1906,

> was recognized as an excellent one but was rather old-fashioned in its equipment and methods. The Gilliland place was recognized as one developing trappy and short-lived apparatus. The Davis and Watts shop was much the same, but not as reliable as the Gilliland Company's. Post & Company were first of all manufacturers of railway appliances and the manufacture of electrical supplies was a side issue with them.[83]

Vail struck a different note when he testified in a legal proceeding in 1908. As he explained it, the key difficulty with the manufacturers was their tendency to "run off in different lines of development."

Scribner strongly reinforced this view twenty years later, when he described the newly launched telephone industry as headed for "design chaos" under the licensing contracts. When the Western Union exchanges were taken over, he wrote, "It became necessary to connect a miscellaneous assortment of apparatus and switchboards of radically different design. . . ." Magneto bells "were all substantially different . . . and furthermore were being changed and modified continuously."[84] The problem, according to these accounts, was one of standardization of equipment, or lack of it, and that, according to Vail and Scribner, led the Bell Company to decide that it must pursue a new manufacturing strategy.

More direct, contemporary evidence suggests that Bell's decision was grounded in different worries than the ones Vail and Scribner addressed. The explicit worries of Bell management in 1880 focused not on standardization—that was a concern of a later era—but on problems arising from the tendency of the licensed manufacturers to deviate from the contract terms regarding pricing, distribution, and submission of reports. Bell could not get the producers to maintain the agreed-upon prices. The company's problem was the exact opposite of that faced by a wartime price-control administration. Bell's manufacturers wanted to (and did) cut prices, not raise them. But Bell found enforcement of price control to be a difficult and frequently futile task. Nor could Bell get the manufacturers to adhere to the contract terms limiting distribution of apparatus to Bell-licensed agents and requiring weekly reports on orders, production, and shipments. These provisions were vital to the defense of the Bell-owned patents, but the manufacturers were more interested in near-term sales than in the future of the Bell patents.

From the perspective of the manufacturers, of course, these problems all looked different. As competitors for the business of Bell-licensed agencies, the manufacturers were deprived by the contracts of a primary competitive weapon—the ability to lower prices. Moreover, the notion that manufacturers would politely stay in regional markets (a notion that the manufacturers seem to have shared to some degree) conflicted with the natural tendency of competitors to make incursions where territories overlapped. With time, these ventures took the producers farther afield, which necessarily invited retaliation and a new round of competition. In short, much of the difficulty lay in the internal contradictions of policies designed to foster competition in some respects but not in others.

Bell, moreover, lacked the administrative wherewithal to enforce its manufacturing agreements. Even gross or obvious violations of the manufacturing contracts, such as the sale of apparatus to "infringers" or the failure to provide timely reports, were awkward for the Bell Company to police. No mechanism of control existed short of the threat of cancellation of the manufacturing license. Such drastic action was unlikely, since it would have inconvenienced all concerned by disrupting established and generally useful business relationships between producers and buyers.

What the Bell Company needed was a new arrangement that would somehow capture the advantages of the licensing contracts without their problems. The advantages were obvious: the contracts had solved the problem of meeting the firm's demand promptly and with high-quality equipment. Bell's ability to stay in touch with and to promote actively the field of telephone technology had been enhanced. This was something with which the firm was increasingly concerned and to which it would devote much more attention and greater resources in the next few years. But weighed against these advantages were the numerous difficulties that the licensees were creating. If Bell was unable to bring them under control, the firm's patents, still its most important property, might be endangered. Repeated efforts to improve reporting and to curb the sales to "infringers" had failed. Price competition had continued, threatening to undo Bell's efforts to stabilize production at minimum standards and to establish a territorial division of labor. The Bell Company, strengthened by the elimination of its major competitor and armed with new financial resources, began to look for a means of eliminating these problems on the manufacturing end of its business.

Elisha Gray (1835–1901) was the guiding technical genius behind the telegraph manufacturing business of Gray and Barton. Gray's claim to having invented the telephone provided the basis for Western Union's entry into the telephone business in late 1877.

Enos M. Barton (1842–1916), was a co-founder of Gray and Barton, the corporate predecessor of Western Electric. Barton was Western Electric secretary (or general manager) in the 1870s, and in 1885 he became president, a position he held until 1909.

General Anson Stager (1825–85), the first president of the Western Electric Manufacturing Company, from 1872 to 1885. Stager, who had commanded the Union army's signal corps during the Civil War, was a long-time employee and executive of the Western Union Telegraph Company. It was after his departure from Western Union in 1881 that he engineered the entry of his manufacturing company into the nascent Bell Telephone System.

The cradle of Western Electric, this Chicago building housed the shops and offices of Gray and Barton in the early 1870s. In 1872 the sign was changed to read "Western Electric Manufacturing Company."

The personnel of Gray and Barton shortly after the firm's removal to Chicago from Cleveland in the early 1870s. Elisha Gray is seated in the front row, center, holding a model of his automatic telegraph printer.

Idealized view of the Western Electric Manufacturing Company's principal factory in Chicago between 1872 and 1884. The Kinzie Street facility accommodated growth in personnel from twenty-five to more than four hundred during its life as Western Electric's headquarters.

In 1877 the top story of the Kinzie Street factory housed workbenches and lathes (shown here) as well as other machinery. A carpenter's shop, insulating room, and foundry were located on the lower floors. Light was provided by the numerous windows and by the skylight located in the center of the room.

Western Union was the nation's largest business corporation in the late 1870s, when it competed with the fledgling Bell Telephone Company for control of the telephone business. After 1873 its headquarters was located at 195 Broadway (the tallest building in the picture), later the site of the corporate headquarters of the American Telephone and Telegraph Company.

The Edison carbon transmitter mounted on a Western Electric–manufactured magneto call box. On the right is the distinctive Western Union "pony crown" receiver, devised by G. M. Phelps, a Western Union "mechanician," in 1878. On the receiver, six permanent magnets bent into horseshoe form were used in place of the single magnet employed in Bell magneto telephones. The switch at the upper left changed the telephone from the stand-by to the talking condition, while the switch at the upper right disconnected the entire apparatus to guard against the demagnetization of the call bell in the event of a thunderstorm.

Unlike contemporary Bell telephone sets in 1879, Western Union (or Gold and Stock) telephones and apparatus were made by one manufacturer and were packaged more as a unit. In this battery-powered call bell, which was operated by a push button *(lower left)*, the Edison transmitter was accompanied by a Gray receiver, which had two diaphragms and a single U-shaped magnet. Battery-powered call bells were adequate for lines a very short distance apart, while magneto ringers were necessary for lines more than a few miles apart.

The Blake variable-contact, battery-powered transmitter became the Bell standard after 1878. Operating on the microphonic principle, the Blake instrument was superior to Thomas Edison's carbon-button transmitter (also a microphone), giving Bell a competitive technological edge in its rivalry with Western Union. Within a year, the electrical shop of Charles Williams, Jr., took the Blake instrument through seven phases of development, reducing it in size and refining its material composition.

An interior view of the Blake transmitter in its seventh phase. It has four binding posts: one for the line, one for the ground, and two for the battery. At the lower left of the box is a cylindrical induction coil. On the back of the door frame is a centrally damped diaphragm, a block of hard carbon mounted on a stiff spring, and a bead of platinum mounted on a light spring. A constant flow of current provided by a battery would be met by the resistance caused by the carbon and platinum making contact at varying degrees of pressure in response to the pulsations of the voice-activated diaphragm. It was the variable resistance of the carbon and platinum electrodes that allowed for the more precise transmission of modulations in the voice than was possible via the magneto transmitter.

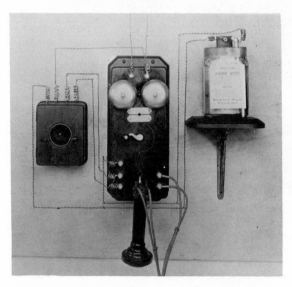

An 1879 subscriber station set consisting of a Blake transmitter, a Williams call bell, a telephone receiver, and a LeClanche battery. The Bell licensee, in order to assemble such a set, would order his telephone receiver and transmitter from the Bell Company (who had purchased them from Williams), his call bell from one of five Bell-licensed manufacturers, and his battery on the open market.

Thomas A. Watson (1854–1934), Alexander Graham Bell's assistant, became the early Bell Company's superintendent of manufacturing in 1877. Through mid-1878 he served as a virtual one-man development operation. By 1880 he was made head of the Electrical Department, the first formal organization for technical development in the firm's history.

Credited with inventing the microphonic principle, Emile Berliner (1851–1929) was employed by the Bell Company in the fall of 1878 to conduct research on the basic telephone instrument. By 1881 he was responsible for performing "original investigation" under the auspices of the Electrical Department.

Francis Blake, Jr. (1850–1913), exemplified the dependence of the early Bell Company on outside sources of invention. His transmitter, which became the Bell standard during 1879, earned him a seat on the company's board of directors.

CHAPTER 4 🐦

Strategies for Vertical Control and the Acquisition of Western Electric, 1880–1882

THE BELL INTERESTS FOUND A SOLUTION to many of their manufacturing problems when they acquired majority control of Western Electric in 1882. With that acquisition, and with the consolidation of some of its other manufacturing licensees into a single corporate entity, the Bell Company restored the controls over pricing and distribution of telephone apparatus that it had once enjoyed in its less formal, unitary arrangement with Charles Williams, Jr., while securing to itself adequate capacity for growing demand. Bell's integration backward into production, however, was not a direct response to its difficulties in enforcing the terms of its 1879 manufacturing licenses. It was a decision taken in the larger contexts of a radically altered market, of shifting constraints on the firm, and of changing perceptions of the company's sphere of business. The decision was an integral part of a larger set of strategies for advancing the business and technological development of telephony on a national scale under Bell Company control. It is necessary to examine these changing contexts and emergent strategies in order to understand why the acquisition took place.

· I ·

Between 1877 and 1880 three critical challenges confronted the Bell patent holders in their attempt to market the telephone: an ever-increasing rate in the growth of demand; ongoing developments in the technology; and heavy competition from a well-financed and technologically sophisticated adversary. The third of these challenges, competition from Western Union, intensified the rates of demand growth and technological change and threatened Bell's claimed right to a monopoly of the market based on its patent claims. We might be tempted to assume that with Western Union's conditional withdrawal from telephony in November 1879, the pressures of demand and technological change diminished, but they did not: they increased.

Demand was stimulated by the very structure of the settlement. A multitude of exchanges developed by Western Union operatives in fifty-five cities were to be taken over by Bell licensees in return for fair compensation.[1] Although Bell also took over 56,000 Gold and Stock telephones, they would soon have to be replaced; and they would have to be replaced by Bell-licensed manufacturers. The new exchanges, in any case, could be expected to grow along with Bell's existing operations. Because of the new harmony between Bell and Western Union, which was to be translated into utility and profit based on the technological complementarity of the telephone and telegraph, demand also could be expected to grow. With Bell exchanges acting as local, voice-grade feeder services for Western Union's long-distance telegraph wires, the value of the telephone to subscribers was greatly enhanced. Aware that its own growth would be largely dependent on the business of its captive feeder service, Western Union agreed to "give assistance and cooperation in extending the use of . . . telephones."[2] William Forbes believed that under its arrangement with Western Union, Bell could expect to double its more than 210 exchanges within a year. Vail's projection of 100,000 new telephone rentals for the fiscal year beginning in March 1880 (in order to more than double the number already in the hands of Bell operating licensees) reflected a clear conviction that the potential for growth was explosive.[3]

In turn, demand, which was concentrating around the ex-change services provided to the mercantile centers of urban places, was having serious technological impact on telephone operations. This was problematic, because while the Bell Company had a substantial measure of control over telephone subscriber equipment, it had practically none over the switching and transmission facilities that connected subscribers to each other. Switchboard design and development occurred almost entirely outside the company—on site at the operating agencies or in the various manufactories. This resulted, as Frank Jewett later explained, in the evolution of "many peculiar and almost fantastic types of switchboards," with relatively little common sharing of intelligence on the increasingly intricate processes of switching. By 1880 no one among the Bell licensees had developed a board that could handle more than seventy-five circuits efficiently, hardly enough to accommodate the rapid growth of the exchange business.[4] Then, too, the proliferating numbers of lines radiating from exchange offices were posing problems in trans-mission. The thickening networks of wire increased the occurrence of induction interference, necessitating more intelligence on the location and spacing of lines and on the materials and techniques employed in their construction.[5]

If the Bell patentees were to exploit fully their potential for growth during the remaining life of their patents, they would have to make sure that their productive capacity stayed abreast of the quickening pace of demand. At the same time, if they were to maximize their potential for growth, they would have to attend to the mounting technical problems of their operating licensees. The common switching and transmission problems of the numerous, scattered, and expanding Bell exchanges required more systematic development of hardware and techniques and more centralized coordination of technical information. Again, the manufacturing capabilities of the company—efficient design and production of equipment—would be crucial to such a process.

All this was well understood by the owners and managers of the National Bell Telephone Company as they extended their planning horizon in the wake of the Western Union settlement. The settlement, by removing the most immediate threat to their busi-ness, allowed Bell officials to turn their thinking from the tactical concerns of competitive warfare to long-term strategies. Realizing that they had fourteen years remaining in the life of the original Bell patents, William Forbes and Theodore Vail looked ahead to 1894. By that time, they hoped, the Bell Company would have secure "control of the field."

"Control of the field," or "occupying the field" (to use the words of early Bell officials), involved strategies that began with a narrow focus on the patents and then widened into elaborate plans for organizational and technological development. The underlying premise, though rarely articulated, was that competition remained a persistent threat to the business. Though dormant, competition could awaken at any time and would, in the long run, reappear in strength. The Bell interests knew just how tenuous their patent monopoly was.[6]

In the short run, the patents were always vulnerable to competition. It is true that the company's patent position was heavily fortified by the transfer of eighty-three Western Union patents to Bell under exclusive license for seventeen years. Bell now held most of the principal claims to the telephone—including the Bell and Gray telephone patents and the Edison, Blake, and Berliner microphonic transmitter patents—and also received rights under the settlement to a variety of signaling and switching (talk-standby) apparatus to be used in conjunction with the subscriber's telephone "set."[7] And yet despite this windfall, Bell was quickly confronted with challenges to its patent base from other sources of invention. At least three compelling patent challenges to the basic telephone surfaced within two years of the Western Union settlement, portending lengthy and expensive litigation.[8] It seemed that tinkerers and their backers everywhere, in quest of a point of entry into the market, were searching for ways to get a leg up on Bell's telephone technology.

In the longer run, the expiration of the Bell telephone patents would leave the company exposed to competition not only from rank newcomers but also from Western Union, which, one might expect, would be well positioned for a vigorous reentry into telephony. Then, too, there might be a future problem in the failure of Bell operating and manufacturing licensees to achieve major breakthroughs in switching technology. Important inventions in

large-capacity switchboards were being developed by the Law Telegraph Company in New York and by Western Electric in Chicago.[9]
Bell could acquiesce in its operating licensees' going to the open
market for switchboards, but if it wished to bring advancing switching technology under its own purview, it would have to either
develop more effective boards of its own or risk negotiations with the
controllers of these critical outside innovations.

Thus, in its narrow, patent-oriented construction, "control
of the field" required that the Bell Company guard its flanks with a
vigilant scanning of inventions that might infringe upon its patents
while pursuing its own program of innovation. Otherwise, it risked
the erosion of its newly won monopoly in the marketplace, either
through the resurgence of competition or (even worse) through
exclusion from some important and fundamental technological
innovation. That is why, as Vail testified in 1908, the company
hoped to "surround the business with all the auxiliary protection
possible, in order to make it indifferent to us whether the [original]
patent was extended or not."[10]

In its broad construction, "control of the field" extended well
beyond the patents to considerations of growth and organization.
The essential issue, as Forbes put it in 1880, was to render the
business "permanent and *independent* of patent rights."[11] The records
and correspondence of the Bell Company from the waning days of
1879 through the first year of the new decade reveal a growing
awareness of the primacy of organization over patents as the key to
enduring success. The occupation of important urban markets with
Bell-licensed exchanges; the development of long-distance telephone transmission independent of Western Union's telegraph
network; and some measure of integration forward into the business
of its operating licensees to ensure their loyalty and to promote
general and uniform standards of business and technical development—these were the functional and structural means by which Bell
hoped to achieve a durable control of the field.

The rapid occupation of important urban exchange markets
was obviously of strategic importance. Bell had learned from its
experience with Western Union that being second in a particular
market posed several problems. Municipalities were often reluctant
to grant more than one company a charter to use streets or viaducts to

string wire; start-up costs in a capital-intensive business were high enough without the added risk of competition; and unless there were large bodies of potential customers far removed from or priced out of existing exchanges, recruitment of new subscribers was likely to be difficult.[12]

It took more imagination to foresee the development of long-distance telephony. Telephone transmission in 1880 relied on noisy, grounded iron-wire circuits and on transmitters incapable of reliable volume beyond a few miles. Yet, despite inadequacy of the contemporary technology to achieve long-distance voice transmission, the Bell Company, in negotiating its agreement with Western Union, fought tenaciously for the right to connect its exchanges with telephonic toll lines.[13] While Bell was explicitly barred by the terms of the settlement from engaging in the interexchange transmission of *telegraphic messages* (by voice or otherwise), it retained the right to transmit *telephonic conversations* over any distance. This concession may not have seemed important to Western Union, because it was generally believed that the main market for long-distance telephony, should it become technically feasible, was for the voice transmission of "general business messages," a *telegraphically* defined market from which the Bell Company was explicitly barred by the settlement.[14] Nevertheless, the Bell Company pressed ahead with technical and organizational experiments in long-distance telephony in 1880.

The strategic implications for long-distance telephony were both defensive and offensive. Once the patents expired, interconnected Bell exchanges would have an entrenched extraterritorial transmission advantage over new entrants into local fields. Once the 1879 patent agreement expired, Bell would no longer have to remain technologically dependent on Western Union for the interexchange transmission of "general business messages." Bell would then be in position to enter the telegraph market on a strong, competitive footing.[15]

Nowhere was the Bell interests' awareness of the tenuous life of their patent monopoly more evident than in their determination to integrate forward into operations. The most important lesson they had learned in their struggle with Western Union was the double necessity for improving their hold on their operating licensees and

for sustaining their equipment and service *in the field* at high standards of performance. On the one hand, agents were too susceptible
to defection in a competitive world to be held in place by patent
licenses if plausible alternatives and superior technologies became
available. On the other hand, agents operating at low standards—
because of insufficient funds, poor management, or weak intelligence—would be too vulnerable to the entry of well-heeled, well-
managed competitors. Therefore, in addition to sustaining their
technological advantage through legal action against patent "infringers," purchase of outside inventions, and internal research and
development, the owners and managers of the Bell Company began
to take control of the operations of telephony.

The most obvious means of control was financial. "Our
company must be ready," Forbes noted in an unsigned memorandum, "to put money in . . . local companies . . . whenever it
is necessary to give . . . prompt and efficient development."[16] But it
would take more than money to bring local companies up to good
standards of operations. Some measure of functional control was
required as well.

Increased functional control over operations had begun
modestly through Bell's attempt to gather and disseminate intelligence on best practice. This was all the company could do with its
limited resources and personnel. In the fall of 1879 the company was
anticipating more direct means of control over its licensees better to
coordinate their technical, business, and financial development in
accordance with the best "general experience." As we have seen, the
heat of competition had already drawn Bell into the financial affairs
of its more vulnerable exchange agencies, usually through pricing
concessions on equipment but sometimes (and more significantly)
through direct infusions of cash in exchange for equity. Financial
investment invited closer monitoring of operations, as in New York,
where the management of local operations was assumed directly by a
Bell official. Elsewhere Bell was increasingly providing advice generally to local companies through its traveling agents and technical
correspondence.[17] By September 1879 this haphazard array of experience was congealing into positive strategy. As Vail explained it to
one licensee, a plan was under way to reorganize the Bell Company
"for the purpose of building lines and exchanges taking an interest in
the exchanges already organized where the parties desire, . . . and

furnishing capital and in other ways pushing the business backed by sufficient capital. This," he argued, "would allow the company to combine in one business the elements that I think will make it strong. . . ."[18]

Expanding on this theme, President Forbes explained to the stockholders at the end of the year that even with its fortified patent position, the company must take tighter control over its growing array of franchises. Doubtless, "branches of the telephone business could [continue] to be parceled out and farmed out to many small companies with some profit," Forbes conceded, but "the work can be done to greater advantage and made of more service to the public if it is controlled and directed by a central organization and brought under a single system. . . ." The idea, as Vail had earlier described it, was to leave management of the exchanges in the hands of local parties, while "giving to each exchange all the benefit of one giant organization with general experience and abundant capital. . . ." In other words, the conduct of operations was to remain decentralized, but the coordination of "general experience" pertaining to operations was to be centralized under the Bell Company's guidance. At the same time, Bell would assume greater responsibility for the financing of a telephone "system."[19]

Thus Forbes and Vail redefined their business. No longer was the Bell Company simply to serve as a patent franchiser arranging for the provision of telephone supplies to independent entities who then placed them into service. The company must now formally undertake integrative tasks pertaining to operations of telephony—tasks that it had already begun to perform out of necessity on a less systematic basis. In order to dictate the balance between local autonomy of operations and more centralized coordination over the general financial business and technical development of telephony, the Bell Company was to assume equity positions in its licensees' firms. In the process, however, Bell would have to overcome its own financial, technical, and administrative constraints.

· II ·

The most severe constraint on the Bell interests' early business development had been their chronic shortage of funds. In a few strokes of the pen, the Western Union settlement had changed all that: the market value of National Bell's common stock soared to at least six times par by the end of 1879. In the heady swell of the company's success at the bargaining table, new investment for telephone enterprise was ripe for the picking. As 1879 drew to a close, a top priority of Bell officials was to secure "financing on a scale not provided for in the arrangements of our company which contemplate only the manufacture and renting of telephones."[20]

First, however, the Bell interests would have to overcome a legal obstacle to deepening their company's pockets. National Bell's nominal capitalization was limited by law to a mere $850,000. If its owners were to capture enough new investment to support their plans for growth and integration forward into their operating licensees, they required both a huge increase in their capitalization and a clearly specified right to hold equity in other corporations. As it was, their current charter left uncertain the National Bell Company's power to hold stock in its agencies.[21] The Commonwealth of Massachusetts was particularly cautious about providing unrestricted sanction to private enterprise. Corporate charters for firms capitalized in excess of $1 million were granted by special acts of the legislature that placed ceilings on capitalization, set limits on ownership in other corporations, and (in keeping with the traditional purpose of incorporation) were contingent upon a demonstration by the grantees that their enterprise was undertaken in the public interest.[22]

And so when the Bell interests approached the legislature for a new charter in the winter of 1879/80, their chief counsel, the renowned patent attorney J. J. Storrow, prepared a brief that defined the Bell enterprise as a public service seeking centralized financial control over the necessarily decentralized operations of far-flung licensees. Wanting no more power than "what the general telegraph law gives," Bell desired that its presumed "power to take stock in other corporations" be made explicit in order to prevent the question from being raised in other states. This would allow Bell, under conditions of tighter control, to continue doing what Storrow argued that it had always done, namely, "to associate local capital and persons with us, that the business of each place may be done by local persons who best understand its needs and the wishes of its inhabitants."[23] In return for the right "to grant exclusive territorial" patent licenses (and for a substantial increase in capitalization), Bell proposed to be "diligent in extending the use of telephones by furnishing them to all." Growth, in Storrow's reasoning, had become a public-service obligation of the patent monopoly.[24]

The Bell interests secured their new charter on 19 May 1880. The American Bell Telephone Company succeeded to the rights and obligations of National Bell and authorized a capitalization of $10 million, or almost twelve times the amount of its corporate predecessor. Granting the right of American Bell "to become a stockholder in . . . other corporations organized . . . or already established for the transaction of telephonic business under its present patents and no others," the legislature limited the stock the firm could take in corporations doing business in Massachusetts to 30 percent.[25] Although Forbes "objected very much to having a 30 percent restriction put in," the charter left ambiguous whether the company could hold higher proportions of capital stock in foreign corporations. Subsequently, the American Bell Company chose to interpret its charter liberally and frequently accepted franchise stock in amounts exceeding 30 percent.[26]

The creation of the American Bell Telephone Company when a high value was placed on Bell telephone stock by investors in the wake of the Western Union settlement brought a decisive end to the firm's weak capital position. American Bell exchanged six shares of its stock for each share of National Bell's for their conservatively estimated market value of $600 per share. Eighty-five hundred additional shares were sold at par ($100), so that the capital stock outstanding was worth $5.95 million, leaving a reserve of more than $4 million in unissued stock. Over the course of the year, the number of stockholders in the firm grew from 380 to 540, indicating an increasing willingness on the part of American Bell's principal owners to go to the market to raise necessary funds.[27]

With the financial constraints on the company alleviated, administrative constraints received more attention. Through the spring of 1878 the daily affairs of the business had been conducted by its three principal active owners—Gardiner Hubbard, Thomas Sanders, and Thomas Watson—with the assistance of no more than a handful of clerks. By 1880 the company was operated by a general manager and as many as twenty specialized personnel, with a payroll in excess of fifty thousand dollars, corresponding to the increasing growth and complexity of the technology and licensee operations. In organizational terms, Bell was being ordered in a more hierarchical, delineated, functional structure. A four-man executive committee led by the company's president, Forbes, made financial decisions, set policy, and conducted high-level correspondence. Vail, as general manager, supervised the firm's day-to-day affairs.[28]

As general manager, Vail was the link between policy and execution. He translated the directives of the owners into operation and in turn advised the president on matters of general financial and strategic concern. Reporting to Vail were Watson, who presided over a newly formed Electrical and Patent Department; Oscar Madden, the former traveling agent, who now became superintendent of agencies; and a variety of other personnel responsible for everything from technical development (including one university-trained engineer to advise operating licensees on underground construction) and legal affairs (five patent attorneys) to bookkeeping, clerical work, and office supervision.[29]

The two new specialized departments, the Agencies Department and the Electrical and Patent Department, reflected the growing formalization of the company's administrative concern with the business and technical functions of its licensees. The traveling agents were given increased responsibility for coordinating company relations with the agencies, for establishing routine forms of accounts and reports, for monitoring local operations, and for disseminating technical information and advice. Gradually, the traveling agents assumed more and more authority over the actual management of operations in some locales, and eventually some were even elected to local boards of directors.[30] Initially, the Electrical and Patent Department's main thrust was patent research. Thomas Lockwood, a technical expert whom Vail had lured away from the American District Telegraph Company of New York in July 1879, was responsible for surveying the entire field of technical developments in telephony in order to evaluate outside inventions either for purchase or infringement warning. At the same time, however, the department was assuming responsibility for a variety of technical functions that had existed on a less formal basis, including the inspection and supervision of manufacturing standards; the collection and preparation of data for technical circulars; the answering of licensees' queries about hardware. Thomas Watson, who would soon leave the firm with a small fortune in search of adventure, was becoming institutionalized.[31]

Thus, in 1880 the Bell telephone enterprise was ceasing to be a closely held venture of limited means and was becoming a large, well-financed, professionally managed corporation projecting long-range plans. Rich in cash, unencumbered by competition, and legally empowered to become a holding company on at least a limited basis, the American Bell Telephone Company was poised for large-scale growth supported by strategies for key market penetration, patent control, technological development, and vertical integration forward.

· III ·

Over the next few years, the American Bell Telephone Company embarked on a series of structural and financial maneuvers that by 1885 gave it an ownership position in most of the firms connected with the telephone industry. It was in this period, with the taking of equity in firms responsible for a wide range of functions along the vertical axis—from development and production to installation and operations on a national scale—that the Bell Telephone System was born.

After 1882 the Bell Company began in earnest its integration forward into the business of its operating licensees by encouraging the consolidation of local agencies into larger, regional firms, in which it then took at least 30 percent interest in trade for exclusive, permanent license contracts.[32] This integration, then, systematized a practice that had begun tentatively on an emergency basis—when Bell had taken equity in faltering but strategically important agencies during the competitive struggles with Western Union—and then had been extended to a few more large urban exchanges. In 1885, with the creation of a wholly owned subsidiary, the American Telephone and Telegraph Company, American Bell took control of the development of interurban, long-distance telephone communications, concluding five years of unsuccessful attempts to build long-distance lines through independent contractors.[33] Preceding these moves forward into the operations end of the business, however, was the Bell Company's decision to consolidate and acquire its sources of supply.

In the record of the company's discussions of strategy, organization, and finance in the winter of 1879/80, considerations of the supply side of the business are conspicuously absent. This can be largely explained by the relative success of the multiple-licensing arrangements that had gone into effect the previous summer. While some problems were already becoming apparent—independent producers did not always abide by Bell's restrictions on pricing and distribution of patented apparatus—the arrangement did accomplish its main purpose of providing telephone equipment in adequate amounts to meet growing demand.[34]

But as the Bell Company's strategic ambitions became more sharply defined, so did the question of its control over its manufacturers. Productive capacity, though adequate in the near term, would now have to be weighed against the company's plans for key market penetration on a national scale. Achieving control over critical technologies for the large-scale development of urban exchanges and long-distance transmission, moreover, required that Bell remain attentive to the entire process of supply—from the conception and development of hardware, through its manufacture and inspection, to its pricing and distribution. The transformation of the Bell enterprise from a contractor of patented hardware into a financier, owner, and coordinator of a national system of telephone exchanges was to be based on a bedrock of backward integration.

The earliest evidence of a plan for backward integration comes not from Bell management but from the pen of Charles Williams, Jr. On 2 March 1880 Williams drafted a scheme to incorporate his manufacturing business and to capitalize it at $300,000. The new company was to retain $180,000 of the stock after selling $60,000 for working capital. The remaining $60,000, or one fifth, was to be given to the National Bell Telephone Company "in consideration of a perpetual license to manufacture under all patents owned or controlled by it. . . ."[35]

Williams's proposal was predicated on his hope of "extending my present business and establishing it on a permanent basis."[36] No doubt cognizant of the Bell Company's plans to embark on an ambitious program of expansion under its patent monopoly, Williams was seeking greater security in his association with the only buyer of his principal line of goods. Like many of the operating agencies, who were also beginning to clamor for perpetual licenses, he wanted some guarantee of permanence; at the same time, he recognized that the small-shop phase of his business had to give way to a more formal effort aimed at financing production on a larger scale and over a longer term.

Replying on behalf of the Bell Company, Oscar Madden, now the assistant general manager, told Williams that his proposal was being considered.[37] No further action was taken; nevertheless, the proposal seems to have stimulated further thinking about the Bell Company's relationship to its producers. Given the logic of Bell's own impending reorganization and plans for expansion, Williams's idea of exchanging stock for a perpetual license illuminated the purchasing power of the Bell patents. The patents, it was clear, could be used as leverage for achieving a solid ownership position in a manufacturing firm.

By July 1880 the idea had taken hold. On 14 July 1880 an unsigned memorandum to Forbes recommended that Bell consolidate and take at least partial interest in its manufacturing licensees. The memorandum urged that "negotiations be opened looking to the consolidation of our manufacturing interests into one Co., with branches located at different points—to give this Co. a perpetual license to manufacture under any and all of our Patents subject of course to such restrictions as will protect us and our licensees in the use of them—And for this perpetual license, to give us a consideration in paid up stock and a voice in the control of the Co." "I know," the writer (probably Vail) added, "we can get . . . at least 30% of the stock, and I think, we can get a much larger percentage."[38]

Thus, within a year of executing contractual agreements with several independent producers, the Bell Company began to consider a limited form of backward integration as a means of bringing its producers under closer control. In retrospect, consolidation of the manufacturers appears consistent with the unitary relationship the company had once enjoyed (and seems to have preferred) with Charles Williams, Jr. The 14 July memorandum was a logical extension of the close control the company had always tried to exercise over the production, pricing, and distribution of patented telephone equipment. Yet, the idea was more radical than it may appear at first. This movement toward vertical integration backward was almost unprecedented. The nearly contemporaneous movement by Standard Oil into the ownership of its sources of supply was based largely on considerations of economies of mass production and finely coordinated distribution of a continuous-process product, economies that were far in excess of any that might be expected in the production and distribution of pieces of hardware. A more relevant example for the Bell management to observe was close at hand—Western Union's relationship with Western Electric. Yet here, whatever advantages of control Western Union may have enjoyed stemmed not so much from its minority holdings in the manufacturer as from the personal tie provided by Western Union vice president and Western Electric president General Anson Stager.[39]

Whatever Bell may have learned from its exposure to the Western Union–Western Electric connection, it discovered nothing about economies of scale in production. As a producer of telephones, Western Electric was big but not, according to Bell accounts, notably efficient. Sanders reported to the stockholders in early 1879 that "the telephones used by the Gold & Stock Tel Co . . . cost many times more than ours and are less effisient [sic]." A few months later, Forbes told one correspondent that Western Electric telephones, though inferior in quality, bore a manufacturing cost nearly *four times* that of those made by Williams.[40] If, in fact, Western Union derived any particular savings in production from its large, captive source of supply, no mention of it has been left to posterity.

Production economies, therefore, were not the main issue. Patents, capacity, innovation, and quality of equipment were. When the Bell Company embarked on a strategy of consolidation and control of its sources of supply, it was in response to these same four concerns that have always underpinned its manufacturing policy. As the new strategy unfolded, it was to the Western Electric

Manufacturing Company that the owners and managers of the Bell Company turned. The reasons for this lie, in part, with the shrewd ambitions of the owners and managers of Western Electric.

· IV ·

Having been effectively excluded from the manufacture of telephone and equipment for the American market by the Bell–Western Union settlement of November 1879, Western Electric had lost a prime opportunity to expand its business. Electrical lighting, traction, and other uses of electrical motors were but in their earliest stages of commercial development, while in the census category of electrical industries telephony was second only to telegraphy.[41] As a market for patented equipment telephony may have already exceeded telegraphy.[42] If Western Electric, then the largest maker of electrical apparatus in the United States, wished to remain in this prime segment of the electrical equipment market, it had to come to terms with the holder of the controlling patents.

As early as the summer of 1879 the officers of Western Electric had become aware of the enormous impact that the impending settlement between Western Union and the Bell Company might have on their business. By July, Anson Stager was convinced that "the Bell Co. or parties outside Western Union" would eventually control the telephone business; and according to Milo Kellogg, a Western Electric engineer, it was necessary to "get in shape . . . for the new work." Getting in shape meant turning to the Bell interests. In New York, D. H. Louderback, the director of Western Electric's new supply branch in that city, was already attempting to solicit business from Vail for office wire, telephone cords, and insulators; and soon he would be seeking rights to manufacture magneto bells under Bell patents.[43]

No encouragement from Vail was forthcoming, however, and when Western Union and Bell came to terms in November, Western Electric was left in a worse position than Stager had probably anticipated. For reasons that are hard to fathom, the Western Union attorneys had made no provision to protect Western Electric's right to produce patented telephone equipment beyond the right to continue making call bells for Gold and Stock exchanges pending their absorption into the Bell fold. And so the manufacturer that had so casually loosened its ties to the fledgling Bell Company in 1878 now found itself, according to one company memoir, "in a pretty bad situation, . . . out of luck in the telephone business."[44] As Enos Barton remembered it, "Our business seemed to have a very poor prospect"; he himself had traveled to Europe in search of foreign business "to take the place of that which we had lost or seemed likely to lose." The Bell Company, for its part, was disinclined to grant any more manufacturing licenses.[45]

By now, Stager had been burned twice in the telephone business. He had advised President Orton of Western Union to reject Hubbard's offer of the Bell patent in 1876 and then had watched helplessly as his pet project of the 1870s, the printing telegraph, was rendered obsolete by the telephone.[46] Now his vice presidency of Western Union turned out to be insufficient to ensure the protection of his manufacturing interests in telephony under the Bell–Western Union contract.

Undaunted, Stager set out to expand his interest in the telephone business in a big way. By 1880 he had become president of several telephone exchanges in Indiana, Illinois, and Iowa. Other Western Electric stockholders had also staked some of their personal wealth in telephony. When the Western Telephone Company was formed as a consolidated licensee of the Bell Company, its constituent parts were nine former Western Union exchanges linked to at least ten stockholders of Western Electric.[47]

While these business ties to the Bell Company were being established on the operating side, Western Electric continued to probe for a way to reenter the production of patented telephone equipment. A face-to-face meeting between Barton and Vail was arranged in May 1880. Together on a boat sailing from Boston to New York, the two general managers gossiped and talked shop. Barton "took a little pains to cultivate his [Vail's] society." Vail, after some grumblings about potential patent infringements, suggested, to Barton's delight, that Western Electric and Charles Williams, Jr., consider consolidating their firms.[48]

From the Bell Company's standpoint, a rapprochement with Western Electric made sense in two important respects. First, there was Western Electric's capacity. Bell's current manufacturing arrangement was adequate for the needs of its operating licensees, but the absorption of Western Union exchanges in the near term and the ambitions of Bell management for the long term would clearly require more productive capacity. Not only was Western Electric large by contemporary standards but it was managed by men who over the years had evinced a desire to grow rapidly. From 1 April 1877 through the same date in 1880, Western Electric had extended its operations to New York and was selling abroad. The company had increased its sales from $273,000 to $877,000 annually.[49] Comparative statistics for 1880 are lacking, but by 1882, Western Electric employed more than 200 people in its Chicago shop alone ("Like Topsey 'it growed,' " remarked Vail) crowded into its 48,000 square feet of floor space, which was twice that of Williams's. Western Electric's sales for March 1882 (just following its acquisition by Bell) were reported at more than $121,000, compared with Williams's $29,000. Its net profit for the fiscal year 1881/82 was more than $78,000, or about one and one-half times that of the Williams and Gilliland shops combined.[50]

Second, there was Western Electric's control of important technology and expertise relating to switching, cable development, and end equipment. In 1880 the manufacturer was in the process of acquiring the rights to Leroy Firman's multiple switchboard, the principle of which promised to become the basis for large-scale switchboard development in urban exchanges. Firman had developed a "multiple"-switching principle while running the Western Electric–affiliated Chicago exchange in 1879. Bell had nothing comparable; its own Boston exchange was poorly equipped for efficient, large-scale switching operations. Western Electric's accomplished electricians Milo Kellogg and Charles Scribner were also known to be working on significant switchboard improvements.[51] In cable technology (which was becoming a *sine qua non* for long-term urban communications even by 1880), Western Electric had developed a lead pipe cable that would become one of three viable alternatives for carrying wire across waterways (and later underground) in the 1880s.[52]

As important and potentially threatening as these developments in switching and cable were to the American Bell Company, its immediate interest in Western Electric technology in 1880 focused on a cluster of telephone transmitter patents owned by one John H. Irwin, an obscure inventor from Morton, Pennsylvania. Irwin, who had also purchased the rights to some telephonic inventions conceived by his neighbor, William L. Voelker, granted the exclusive use of his and Voelker's patent claims for "speaking telephones," "telephone transmitters," "acoustic telegraphs," and "electric telephones" to Western Electric in exchange for a very generous schedule of inventor's royalties on 20 January 1880. The patents on telephone transmitters were especially significant, since they would soon be wielded as a threat to the Blake transmitter, the Bell Company's standard since 1878.[53]

Vail perceived the threat immediately. One week after the Irwin–Western Electric agreement, he invited Irwin to consider an outright sale of his patents to the Bell Company. Although Vail was not prepared to admit to any validity in Irwin's claim on "the microphonic principle," nevertheless, the company was willing to negotiate a settlement. Another letter followed pressing Irwin "to the transfer of your claims to this company," and after two more months of correspondence, Irwin was ready to meet. The conference, however, came to naught.[54]

Vail's concern about the Irwin patents played right into the hands of Western Electric. A letter from Barton to Kellogg provides a glimpse of the wider scheme in which Irwin had been given a role. Details are lacking, but apparently the Irwin-Voelker patents had become tied to a plan of the manufacturer's "parent company," Western Union, to use the patents as leverage for gaining more favorable terms in the impending transfer of some of its telephone exchanges to Bell under the terms of the 1879 settlement. Although Barton was somewhat distressed by Western Union's manipulation of the Irwin-Voelker patents (he was angry enough over Western Union's wholesale divestiture of its patents in the 1879 settlement), Western Union's involvement served to reinforce the value of the Irwin-Voelker patents to Bell.[55]

By the end of May, Western Electric had turned the Irwin-Voelker patents to its own advantage. Barton's meeting with Vail revealed the depth of the latter's interest in acquiring *all* potentially important patents affecting telephony. In haggling over the Irwin-

Voelker claims, Vail disclosed the Bell Company's fear that by coming to terms on a single set of patents with Western Electric, it risked an ongoing and expensive series of such transactions. Barton seized upon this opportunity to offer Vail a chance to negotiate an agreement that would ensure for the future the Bell Company's control of Western Electric's entire array of telephonic patents, including those for large-scale switchboards, in which Vail had expressed a special interest. The Irwin-Voelker patents were, in short, the tip of the iceberg. Vail, as Barton related to Kellogg,

> inquired what we proposed to do in the matter of switches and other patents that we might find in our own way hereafter. In other words [he] said if we make a trade with you now how often shall we have to repeat it with the Western Electric. I said I thought so far as telephones went we would put the future right into this trade. I thought said I you were ambitious to control the telephone patents, but you don't have an idea of acquiring the collateral things such as exchange apparatus, etc. do you? Most assuredly we do, says he.[56]

The Irwin-Voelker patents became both a stick and a carrot in the hands of Western Electric as it pressed the Bell Company for a manufacturing license. Bell, in its desire to gain control of all potentially important patents, especially those that might compete with its own claims, might have been indifferent to the means by which such patents came under its control (whether by purchase or by license) had the purveyor been an occasional inventor. But Western Electric had the potential to generate (or acquire) technical innovations on an ongoing basis. Thus, when Barton offered Vail use of the Irwin-Voelker patents under a sublicense for twenty-five thousand dollars in cash plus a royalty of one dollar per annum on each transmitter made under the patents, Vail (predictably) refused. When Vail refused, Western Electric, with Irwin's cooperation, sued the Bell Company for patent infringement, claiming priority of the Irwin-Voelker transmitters over anything Bell had under its control. Each side, it seems, was holding out for larger stakes.[57]

It was a long way from the spring of 1880, when Vail and Barton began to discuss their mutual business interests, to the spring of 1881, when the two companies set in motion negotiations for the exchange of a Bell manufacturing license for Western Electric stock.

In the interim, two events occurred—one initiated by Western Electric, the other entirely fortuitous—that proved decisive. The first event was the acquisition of the Indianapolis Telephone Company by Western Electric. Already attractive because of its large capacity and technical resources, Western Electric now became, by virtue of its control over Gilliland's shop, almost indispensable to Bell. The second event was the takeover of Western Union by a new owner, which made Western Electric, long-time captive manufacturer to the telegraph giant, more readily available for acquisition.

Sometime before November 1879, Western Electric had entered into negotiations for an interest in the Indianapolis Telephone Company, now known as the E. T. Gilliland Company, the American Bell Company's most innovative and important apparatus producer. Whether Gilliland himself was in control of the firm is not certain, but Gilliland was becoming less a tinkerer and more an entrepreneur. Having acquired an interest in several telephone exchanges, Gilliland was broadening his business activities and moving away from his former, single-minded loyalty to the Bell interests.[58]

In January 1880 Gilliland entered into close association with Western Electric when the latter's Chicago agents supported him in his consolidation of the local Western Union and Bell agencies in Indianapolis. Details of the precise nature of the growing relationship between Gilliland and Western Electric are obscure, but in March 1881 Western purchased a 61 percent share in his factory,[59] thereby acquiring control of what may well have been (if we can give credence to the advertising) "the largest establishment in the world making an exclusive specialty of telephonic apparatus."[60]

There was, in other words, more than one way to skin a cat. Having lost its Western Union telephone business, and barred from producing apparatus under the Bell patents, Western Electric reentered telephony via horizontal integration with a key Bell manufacturing licensee. The acquisition of Gilliland now made it almost imperative for the Bell Company to come to terms with Western Electric on the matter of a manufacturing license. Western Electric, after all, could cancel Gilliland's on six months' notice.[61] The American Bell Company, however, did not have to be brought to heel by real or implied threats. Since at least July 1880, Bell officials had been considering granting a permanent, or "perpetual," license to a manufacturer in exchange for a third of its equity. But if Bell had designs on Western Electric, not only would it have had to persuade all of the latter's minority stockholders to sell their shares but it would have had to be content to become a minority stockholder in the face of nearly monolithic control by Western Union of the remaining two thirds.[62] Under such circumstances, a 30-plus percent share in a company might not carry with it much influence should conflicts of interest arise between the demands of telephony and telegraphy.

In the spring of 1881, coincidental to Western Electric's acquisition of Gilliland, an unexpected opportunity appeared for Bell to take Western Electric from Western Union. Jay Gould, whose mounting assault on Western Union in 1879 had already played a role in that corporation's withdrawal from the telephone business, inadvertently delivered Western Electric into the hands of the American Bell Company. In January 1881 Gould stunned the financial community when he suddenly and without fanfare announced that he had seized control of Western Union from William H. Vanderbilt. Gould's formation of competing telegraph companies and his intensive rumor-mongering on Wall Street had sent the price of Western Union's stock tumbling through 1880. Secretly, through agents, Gould bought up the depreciated shares as they were dumped on the market. After he emerged as the principal owner of Western Union in January 1881, he cleaned house. Many of Western Union's top managers and directors were quickly replaced with Gould's own people.[63]

Anson Stager was one of the casualties. With his ties to Western Union severed and with his modest fortune sunk in telephone exchanges and electrical manufacturing, Stager's concern for an alliance with the Bell Company grew. The fate of Western Union also caused alarm among Western Electric's other individual stockholders (many of whom were also employees), who feared the possible effects on their business of any further manipulations by a man who was widely regarded as an unethical financier interested only in milking the businesses he acquired. Under the circumstances, control by the Bell interests must have seemed far more appealing.[64]

In the meantime, the Bell Company was outlining a plan for the transfer to it of Western Electric stock and the Irwin-Voelker patents in exchange for a permanent manufacturing license. Just prior to a meeting between Forbes and Stager, a memorandum (probably prepared as an agenda) proposed the following:

> A.B.T. Co. to give a permanent license to the W.E. Co. to manufacture under all patents now or hereafter held by the A.B.T. Co.—Telephone apparatus—except telephones—receiving therefor $50,000 stock of the W.E. Co. (additional stock to be issued for this purpose), upon terms as favorable as to royalties as shall be granted to any other Company.

> W.E. Co. to agree to license only the licensees of the A.B.T. Co. to use telephone apparatus covered by its patents, or this license.

> W.E. Co. to keep out of telephone business except under license of A.B.T. Co. and to assign to A.B.T. Co. all patents for telephones, the A.B.T. Co. paying, if they use such, any royalties called for by the contracts of the W.E. Co. with the special provision in the case of the Irwin Voelker claims, that in case it is established by the courts that those patents own the exclusive right to use the battery telephone the A.B.T. Co. shall pay the $10,000 already advanced to Irwin and the other sums stipulated for by the contract between the W.E. Co. & Irwin—otherwise the Western Electric Co. shall discharge its own obligations to Irwin.

> If the inventions of Irwin or Voelker patented or applied for shall be found legally to entitle them to a broad claim covering all microphones—or if they shall be found to cover those features of the Blake transmitter which are substantially necessary to its successful practical use then the A.B.T. Co. are to have the exclusive right and license to use them upon the royalty named in the Irwin license of January 1880.[65]

This modest plan would have made Western Electric one of the Bell Company's several apparatus-manufacturing licensees in exchange for what would have amounted to only about 14.3 percent of the manufacturer's stock.[66] To be sure, by such an agreement Bell would put an end to its anxiety over the Irwin-Voelker patents and would bring Western Electric's considerable technological achievements, present and expected, under its control. For this, it would grant a permanent license. But with Stager's departure from Western Union, discussions between Bell and Western Electric mushroomed into a wider consideration of the consolidation of Bell's manufacturing interests into one large concern. With Western Electric already in control of Gilliland and the Detroit Electrical Works (the small manufacturer licensed to make apparatus for Bell's Michigan agency),[67] a further consolidation with Charles Williams, Jr., made sense. It would require Stager's cooperation.

Following the meeting between Forbes and Stager in April, Vail went to Chicago. On 1 June he reported to Forbes, who was vacationing on the West Coast, on a proposition for a "Consolidated Mfg. Co." arising from the merger of Gilliland and Charles Williams into Western Electric. When Forbes returned to Boston, he wrote to Stager requesting him to "secure for us upon reasonable terms enough of the W. Electric stock to give us a majority interest with what we shall receive for franchises." Although Stager responded that he would do his best to secure from Western Union "approval . . . to the proposed scheme of consolidation," he protested that consolidation, on the one hand, and Bell's desire for majority control, on the other, were distinct and separate issues.[68]

As it happened, when Norvin Green, president of Western Union, was convinced that the hostility of Stager and the other Western Electric stockholders had stripped his firm of its virtual control of the manufacturer, he was content to sell Western Union's one-third interest to the American Bell Company on 5 July for $150,000, or $50,000 more than its par value. Stager was satisfied with this outcome, and if he had any misgivings about giving Bell control of Western Electric, they soon vanished. Over the next few months, he used his good offices to secure for the Bell Company enough additional stock, including some of his own holdings, to bring Western Electric under Bell's control. On 23 July Charles Williams came aboard with an offer to sell his firm to the American Bell Company for $120,000 in return for "cash or . . . stock of the Consolidated Manufacturing Company to be organized from the Western Electric . . . , Gilliland . . . & C. Williams, Jr's Manufactory, capital stock to be one million dollars. . . ."[69] It would take several more months to iron out the wrinkles.

On 6 February 1882, amidst a complex series of stock transactions among the owners of Bell, Western Electric, Gilliland, and
Charles Williams, Jr., a manufacturing contract was signed. In it a
newly incorporated Western Electric Company of Illinois received
permanent and exclusive rights to manufacture telephones and
telephone apparatus under the Bell patents. For this license, Bell
took a controlling interest in Western Electric under a form of trust
agreement—a wary concession to the 30 percent restriction in its
charter—until an 1883 amendment to its charter uncapped the
proportions of stock it could take in its operating and manufacturing
licensees. Western Electric was recapitalized at $1,000,000—$600,000
for the purchase of the assets of the former Western Electric Manufacturing Company; $80,000 for the rights, patents, and assets of
Gilliland; $120,000 for the assets of Charles Williams, Jr.; and the
balance for the license from American Bell.[70]

In April, by which time Bell held 53 percent of Western
Electric's stock, the following announcement appeared in the annual
report:

> To obtain a permanent interest in the manufacture of telephones
> and apparatus, as well as to ensure the highest standards in the
> same, we have bought the plant and business of Charles Williams,
> Jr., of Boston, and an interest in the Western Electric Manu-
> facturing Company of Illinois, and propose to merge the two in a
> consolidated company, which will avail of the good will, business,
> and patents owned by that company, as well as our own, and secure
> an economical management for the whole of our manufacturing
> interests. We expect to make this an important part of our
> business.[71]

What Bell had acquired, in other words, was a greatly expanded
production capacity and an enlarged patent portfolio, along with the
opportunity to manage more efficiently its source of supply. The
company now had the means, through corporate consolidation of its
manufacturing interests, to eliminate the pricing and distribution
problems inherent in the 1879 license contracts. By mid-1883 the
consolidation of the firms of Charles Williams, Jr., and E. T.
Gilliland into Western Electric was complete. The licensees of Bell's
other manufacturing agents were terminated.[72] The vertical integration of the Bell Telephone System was under way.

CHAPTER 5 🍃

The Aftermath: Consequences of the Acquisition, 1882–1915

ACQUISITION IS ONE MATTER, INTEGRATION ANOTHER. The historic process by which Western Electric enlarged upon its role as manufacturer to assume the manifold technical functions of research laboratory, equipment designer, central office installer, and general purchasing and supply agent for the Bell Telephone System was a gradual one. In the immediate aftermath of the acquisition, the integration of Western Electric was defined mainly by the exclusivity of its license contract, the controlling equity interest held in it by the American Bell Company, and by the horizontal combination of its factories with those of E. T. Gilliland and Charles Williams, Jr. In structural terms, the connection between Western and Bell was far more legal and financial than technological and economic. The business interests of Bell and Western Electric were mutual but not precisely congruent. Not until the twentieth century—largely in response to increasing technological constraints on the development of telephony—did Western Electric become, in the words of one of its latter-day presidents, wholly "associated with the telephone industry . . . responsible to its needs . . . controlled by the parent company in the interest of the telephone industry rather than in the interest of the manufacturer."[1]

Just how Western Electric, acquired by Bell principally for its capacity and patents, evolved into a functionally integrated sub-sidiary of the Bell Telephone System is worth some consideration. Therefore, we shall look briefly and impressionistically at the process, if only to help distinguish the well-known effect of the acquisition from its motives and causes.

· I ·

The manufacturing license granted to Western Electric on 6 February 1882 was designed in accordance with the same basic policies and objectives as those that had governed all the Bell Company's manu-facturing arrangements since 1877. The policy governing the pro-duction and distribution of the telephone receiver and transmitter was unchanged. Telephones—that is, "all instruments employed for the electrical transmission of articulate speech"—were to be made by a single producer for the Bell Company, which would then (after testing) lease them to its operating licensees.[2] Policies governing the quality, pricing, and distribution of apparatus (or "telephone appli-ances")—now defined more broadly to include "calls, switches, switchboards, annunciators, exchange furniture, and other appa-ratus and devices for use on or for telephone lines"—were also unchanged. Uniform prices for all apparatus were to be controlled by the Bell Company, and apparatus was to be sold to the Bell Company and its licensees only.[3]

Other provisions of the 1882 contract have a familiar ring. Bell's abiding concern with patents was covered in a provision allowing it to acquire any improvements or inventions in telephony then owned or subsequently developed by the manufacturer. This allowed Bell to purchase immediately Western Electric's claims to the coveted Irwin-Voelker patents and secured Western Electric's important developments in switching and cable technology. For the future, this would guarantee Bell's access to the telephonic inno-vations generated by Western Electric employees, who were certain to be a source of useful invention.[4] Bell also maintained its insistence on quality control by requiring submission of Western Electric products for inspection and testing.[5]

The only innovative, and ultimately controlling, feature of the 1882 agreement was a small provision relating to excess capacity. As a formal requirement of production it was new, but it reflected Bell's abiding concern for high volume to support rapid market expansion. Section 4(e) enjoined Western Electric to *"promptly manufacture and supply all such apparatus* as it is hereby authorized to supply, and *of such patterns and styles* embodying such inventions and improvements *as may be ordered* by the parties to whom it is authorized to dispose of the same and will provide itself with facilities for making such apparatus *in such quantities as the course of business may demand."*[6]

It was this provision for excess capacity, more than any other part of the license contract, that laid the groundwork for the integration of Western Electric into the Bell Telephone System. Section 4(e) had the effect, as Enos Barton noted a quarter-century later, of making "it necessary for us to neglect other branches of the business in order to take care of our obligations under this clause of the contract."[7] Or, as Henry B. Thayer, another senior Western Electric official, put it, "We have had to provide a plant considerably in excess of what a manufacturer who is not under obligation would consider prudent. . . . We have also had to sacrifice other lines of manufacture."[8] Thus, in addition to its improved control over the pricing and distribution of hardware through its power of ownership of a manufacturer, Bell now had, for the first time, control over quantity,[9] a form of control to which no independent producer would ordinarily consent.

From Western Electric's standpoint, this requirement for telephone production capacity was no great burden in the long run. But in the years just following the acquisition, Western Electric, despite its close connection with Bell, took a broad view of its business. As "the principal electrical manufacturer of the country," it had been engaged in a variety of technical applications for several markets, making dynamos, printing instruments, fire alarms, and other electrical devices in addition to telegraph and telephone equipment. After the acquisition, Western Electric, by necessity, gave the telephone business "the first place in our efforts," Barton said years later, "but we had never been makers of telephone apparatus exclusively." From the 1880s into the early twentieth century, the company had its eyes on two other major emerging markets for electrical applications: electric lighting and power machinery.[10]

In January 1882 Barton informed Vail of Western Electric's work in vacuum lamps, arc lamps, and dynamo machines and of his intention to devote some of his best personnel and the Gilliland shop to the development of these new applications. Vail responded with qualified enthusiasm. Although he was willing to see Western Electric try to capture "some of the Electric Light business, which promises to be very large," nevertheless he thought that Gilliland's time (in particular) was too valuable to be distracted from telephony.[11] Development went ahead, however, and by 1885 Western Electric was pursuing a modest business in electric lighting and operating at least two midwestern power plants in an attempt to establish a market for electric lamps and dynamos.[12] The evidence is scattered and thin, but it seems clear that Western Electric continued to devote a substantial portion of its capital to the supply of miscellaneous electrical apparatus, such as incandescent and arc lamps, sockets, switches, and line materials, along with the production of power apparatus—machinery and accessories.[13]

Western Electric's continued activity in the power apparatus business was the subject of serious concern by 1907. Unlike the electrical supply business, which, in its distribution patterns, had begun to contribute to the marketing of telephones, the power apparatus business had "only one point of contact with the telephone business, that being the manufacture of telephone motors used with central energy switchboards." Moreover, as a secondary line of business, power apparatus had been neither well enough managed nor sufficiently capitalized to compete profitably with the products of General Electric and Westinghouse. Contrary to Barton's hope of continuing to make at least small-sized turbine engines, Western Electric vice president Henry Thayer, supported by comptroller Charles G. DuBois, persuaded Vail (who had just become president of AT&T, the parent company of the Bell System since 1900) to liquidate the manufacturer's power business. According to Thayer, Western Electric's "real business is the telephone business," and there was no longer any reason to employ any of its capital in anything other than the development of manufacturing "more intimately associated with our business."[14] In 1910, after lengthy negotiations, Western Electric sold its power apparatus facilities to the General Electric Corporation.[15]

This narrowing of the scope of Western Electric's mission was in keeping with the tremendous growth of telephony after the expiration of the original Bell patents in 1894. The advent of competition stimulated a far sharper rise in demand than even Bell officials had anticipated.[16] In 1894 there had been 240,000 telephones, or less than 4 telephones per 1,000 population, in service.

Competition, combined with a general rise in American living standards, stimulated a surge in demand for telephones in rural areas and in private residences, in addition to an intensifying demand for service in the more traditional urban and business markets. Between 1895 and 1907 (by which time the Bell and non-Bell telephones in the United States were about equal in number) the number of Bell instruments placed in service increased at an average rate of 22.44 percent per annum. By 1909 the number of telephones in service in the United States had grown to nearly seven million. Of this number, less than 60 percent belonged to the Bell Telephone System, but that 60 percent was enough to tax even Western Electric's rapidly expanding capacity.[17]

Even though it was thus constrained from broadening its activities in nontelephonic fields, Western Electric flourished. By the end of the nineteenth century the company employed 8,500 people and reported annual sales of more than $16 million (including its business in electrical supplies and power apparatus). In 1903 construction of the great Hawthorne Works in Chicago was begun, and the expansion of telephony was so rapid that the new facility almost immediately required enlargement. Meanwhile, the company was opening a chain of distribution and repair centers around the country in a new role as general purchasing agent for the Bell operating companies.[18] In 1912, by which time AT&T, as the parent company of the Bell System, had gathered in more than 80 percent of Western Electric's common stock, the "captive supplier" of the Bell System was the third largest electrical manufacturing firm in the world, reporting over $66 million in sales. Its business extended to some eight manufacturing plants overseas and one in Canada, but its main market and greatest prospects for growth remained the domestic Bell System. Even though its net return on investment was a comparatively modest 7.8 percent (reflecting a policy of keeping prices for the Bell System at competitive, or "fair," levels), Western Electric enjoyed a secure position in a booming business.[19] On the eve of World War I the number of telephones in service in the United States was more than 11 million, or nearly 11 instruments per 100 population. Bell's share of that was more than 6.5 million telephones, and there was no end of growth in sight.[20] Concomitantly, the intensification of telephone use in urban areas and the achievement, in 1915, of coast-to-coast long-distance transmission required the production of ever-growing and more complex systems of switching and more numerous and refined instruments, apparatus, cable, and wire.

Western Electric's fate as a telephone manufacturer had thus been sealed by the 1882 contract's requirements that it devote its capacity to the priorities of the telephone business. This demand of the Bell System on Western Electric, and Western Electric's response over time to the expanding production needs, strictly in terms of quantity, became a controlling factor in Bell's backward integration. The arrangement for excess capacity at the time of the acquisition, along with the Bell Company's power to exercise its prerogative over Western Electric's choices of production, enabled Bell to hold its own in the expanding and hotly competitive markets of the early twentieth century.

· II ·

Unanticipated at the time of acquisition was the process by which Western Electric eventually became the technologically integrated entity modern AT&T managers have come to take for granted. During the 1880s there were no extraordinary linkages between Western Electric production of hardware and the entities that assembled the hardware into operational systems for the ongoing use of consumers. The Bell operating companies ordered equipment; Western made it and shipped it. In the case of more complex apparatus, such as switchboards, producer and buyer often collaborated on design of specifications, but this would have been possible under wholly unintegrated circumstances. Equipment was produced on order, ad hoc, without much concern over general standards of design.

The engineering of the central exchange office, from the design of the large, multiple switchboard to its installation, became the chief operational constraint on the development of the Bell System by the end of the decade. Engineering the central exchange inevitably drew Western Electric into more regular contact with the operating companies on a continuing basis. As early as 1881, Emile Berliner hailed the acquisition and consolidation of manufacturers as an opportunity to "insure uniformity of systems,"[21] but this was only gradually realized. Early attempts to share intelligence on the common business and technical problems of the exchanges were formalized in the meetings of the National Telephone Exchange Association, which began in 1880 and lasted through the decade. More specialized conferences on switching and cable development, to which key officers and engineers of operating companies and Western Electric were invited, were sponsored by American Bell (occasioned by pressure from its long-distance subsidiary, AT&T). These were occasional but important attempts to coordinate design and production with exchange operations from a systemic point of view.

It was around the central exchange switchboard that the standardization of hardware and practice on a systemwide basis evolved most significantly. In this context standardization was more than the simple reduction of hardware to a set of specifications; it involved the integration of experience in a dynamic setting of technological evolution. More than any other aspect of telephony, the function of switching required constant development, adjusting, and alteration in accordance with the growing and shifting requirements of communications traffic.

As time went on, switching required ever closer coordination between the producers and the users of switchboards. In 1880, exchanges were still small enough in scale that relatively few small, relatively simple switchboards connected by trunk lines could serve even the largest exchanges. But the concentrated growth in demand for telephones in large urban centers required the manufacture of sophisticated systems that by the late 1880s theoretically could accommodate up to ten thousand subscribers each. Such scale

brought complexity and increasing pressure to anticipate the problems of use in the earliest stages of production.

In large-scale switching, establishing the operational requirements for coordination of human labor and the mechanical
processes of the apparatus became crucial at the initial stages of
apparatus design. The tolerances for mechanical error or deficiency
became more stringent, requiring closer attention to precision in
design and production. Moreover, the diseconomies of switching—
in both the proliferation of connections and the complexity of
construction as boards grew larger—were such that while in 1880 a
50–line "standard" board cost $150 to install, a 10,000-line board
planned for the New York City exchange in 1887 was projected by
Thomas Lockwood to cost from $700,000 to $900,000, or as much as
an average ocean steamship. As Lockwood explained, much closer
cooperation was required between the maker of switchboards and the
operators of exchanges in order to design high-capacity switchboards
at a reasonable cost and without loss of mechanical efficiency.[22]

Lockwood's point was addressed by Enos Barton, whose
remarks at the first Bell System switchboard conference revealed the
degree to which Western Electric regarded its relationship to its Bell
customers in conventional, nonsystematic terms. Barton (who had
become Western's president in 1885) argued that the burden for
design for central office systems lay not with the manufacturer but
with the operating companies. In other words, "The manufacturers
of switchboards give their best advice . . . , but they are entirely
agreeable to making switchboards with any arrangement of the parts
that customers want and will pay for. The ultimate responsibility in
that respect must necessarily rest upon the purchasers . . . , because
the manufacturers are only working to please them."[23]

Within a few years this attitude was transformed. The high
cost and technological imperatives of complex switching operations,
including the problem of integrating efficiently the operations of the
machine with the capacity of human labor, necessitated more
systematic planning in the design of central office apparatus. Otherwise, it would not be possible to balance the unique characteristics of
particular exchanges with the best, most cost-efficient techniques
and hardware derived from general experience. This meant that the
manufacturer must follow the switchboard through its functional

implementation, the results of which were ongoing and constantly changing tests of the design and production of apparatus. Increasingly, Western technicians entered the offices of the operating companies to install, modify, and repair the apparatus as might be required until, by the mid-1890s, it became standard procedure. Thus, over time, the manufacturer became responsible for the entire process of switchboard design, manufacture, installation, and even maintenance. And while every exchange office was unique, and while switchboards were modified to suit the peculiarities of buildings and local traffic conditions, successful adaptations in the design and operations of central exchange facilities were generalized throughout the Bell System via the Western Electric engineer.[24]

As the problems of switching involved Western Electric more directly in the technical and operating concerns of the operating companies, problems in transmission drove it back into areas of fundamental research that went far deeper into pure science than did the customary engineering activities of electricians and mechanics. The transformation of telephony from a "mechanical art" into a science paralleled the applied development of electricity in other industries. It occurred primarily because the technical requirements of long-distance transmission stood just beyond the bounds of contemporary, accessible scientific knowledge. In 1900 the practical development of the loading coil, which made possible the significant reduction of attenuation in long aerial wires and underground cables, was the successful outcome of the purchase of a patent from a Columbia University scientist and internal research. At American Bell, George Campbell derived a commercially viable application for loading transmission lines from theoretical physics. The impact of this technological breakthrough on Western Electric was twofold. In the first place, loading coils required unprecedented tolerances in their manufacture, suggesting closer coordination of research, equipment design, and the production process. This, then, set in train the growth of a permanent scientific research function at Western. Campbell was transferred there in 1907, and within four years a research branch was established at Western Electric in response to yet another scientifically rigorous problem in transmission, the development of an electronic repeater, which led to the first successful transcontinental telephone call in 1915. Thus, the kind of research function that had originated in the parent company with the hiring of Emile Berliner in 1878 was now integrated directly into production.[25]

· III ·

By the twentieth century the growing technological complexity of telephony generated a need for more rigorous requirements for the standardization of apparatus at higher levels of quality. Although Bell had always insisted on good-quality equipment from its manufacturing licensees, in the early 1880s it had not been very concerned about the rigid standardization of hardware. By the 1890s, however, the technical demands of long-distance transmission and the increasingly interactive and technically interdependent nature of the large-scale exchange required the integration of more sophisticated equipment made with costlier materials, more precisely calibrated mechanical features, and more finely measured electrical impulses.

Major innovations, such as the development of the copper-wire metallic circuit in the 1880s and the loading coil in the 1900s, had fundamental effects throughout the system on the design of transmission, switching, and subscriber station equipment. A long-distance metallic circuit, for example, could not be used in conjunction with switchboards designed for iron wire. Switchboards designed for compatibility with long-distance metallic circuits required, in turn, compatible local wires, which then required compatible end equipment. Loaded lines brought with them a set of highly sophisticated technical problems with respect to their compatibility or efficiency when used with other parts of the existing wire plant.[26] Even such a mundane process as the introduction of a common battery switchboard into exchange offices—thereby dispensing with the inconvenience and cost of maintaining batteries in the subscriber's telephone set, eliminating the problems of variable local battery efficiency, and aggregating in the hands of the operator more control over the completion of calls—exemplifies the remarkable degree to which the integration of compatible systems of hardware had become necessary. As one expert on the technological history has explained it, "With magneto boards, a fairly relaxed view of voltage, magneto output, line quality, and a variety of transmitters and receivers could be tolerated. But with a common battery system, the tolerance for quality operation was narrower, and every component had to meet its specifications or it would destroy the chain of interactions which were necessary for proper overall functioning."[27]

As the scale and complexity of telephone technology increased, so did the need for rigorous attention to quality standards and creative concern for standardized practice. Initially, manufacturing standards were not easy to enforce. At the time of the acquisition, the American Bell Company retained the responsibility for testing and inspecting Western Electric's products and was not always happy with what it found. Apart from Vail's own observation that the manufacturer was not very well managed, he was chagrined by reports that Western Electric's call bells and batteries were deemed substandard by many of the operating licensees. The Gilliland shop, moreover, while presumably reducing costs through its mass-production techniques, was cranking out bells and switchboards that were considered less durable than those of the more conventional factories. Vail's concern that the monopoly position of Western Electric might have removed some of the incentive to sustain high standards of production was reflected in a toughly worded letter to Enos Barton in the summer of 1882. "There is one thing," Vail warned, "that the W.E. Co. cannot afford to do, and that is, conduct its business upon the basis, that it has an *absolute* monopoly of the telephonic apparatus manufacturing business. This would be exceedingly damaging from every point of view."[28]

Bell's management felt so strongly about this that when it terminated Post and Company's manufacturing license, in accordance with its commitment to Western Electric in the 1882 contract, arrangements were made for the creation of a new company, the Standard Electrical Works, a Post and Company subsidiary, in which American Bell took a one-third interest in capital stock.[29] In 1884, within a year after Gilliland and Williams had completed the sale of their shops to Western Electric, their plants were closed, and their operations concentrated with Western Electric's in its New York shop and in a large, new factory on Clinton Street in Chicago, where the standards of production were expected to improve under more unified management. Complaints about the quality of Western

Electric apparatus diminished by the 1890s, when concern over standards was subsumed by a heightened concern for standardization in response to the more complex technical requirements for apparatus.

The process of hardware standardization began with the telephone itself just following the acquisition. Up to that time, according to one Western Electric executive, "we had never held our product to some uniform standard with which it had to agree until we undertook telephone and transmitter work for the Bell." Initially, given the "haphazard" nature of craft production, it was difficult for the Western Electric shop superintendent to get even half his telephones accepted by the American Bell inspectors, who insisted on a high degree of uniformity in the basic instrument. This problem was overcome soon enough, but for the rest of the decade, as the number of employees at the Clinton Street factory swelled to one thousand, there was little change in the traditional, small-shop methods of production.[30]

The principal constraint on the uniform standardization of hardware was the nineteenth-century "contract system." As H. F. Albright described this system at Western Electric, "Each foreman contracted with the Company to turn out a certain amount of apparatus at an agreed figure. The Company furnished the stock and the machinery, while the contracting foreman secured the workmen and got the job done—of course, as cheaply as possible, since every cent he saved went into his own pocket." The contract system also retarded the control over work desired by the company's general management, who wanted to exercise closer scrutiny over the processes of production, with inspections at each stage of production. It became necessary, moreover, to rationalize the increasingly large number of specialized tasks that were arising in response to the size and complexity of the plant's operations and to achieve better control over the flows of materials and work from one department to the next. Not until 1897 was the power of the foreman to control the processes and quality of work in his own department subordinated to the authority of a new cadre of management inspectors. The arbitrary authority of the inspectors in turn was reduced in 1899 by the introduction of piece-part specification drawings, to which the making of all hardware thereafter had to conform.[31]

All this made possible the transfer to Western Electric from AT&T of the responsibility for final equipment inspection and testing in 1903. But there remained the larger problem of coordinating specifications and practices of the manufacturer, the operating companies, and the long-distance operations of the parent company, all of which had long had their own engineering functions.

In 1907 a major reorganization of engineering activities in the Bell System centralized research and development policies and practices under John J. Carty, the new chief engineer at AT&T. Carty further defined Western Electric's responsibilities for standardization of equipment as he sought to reduce what he considered an "excessive and uneconomical diversity of types of apparatus and methods" throughout the system. To achieve "greater economy and efficiency," Carty recommended "substantial changes." The task of "developing the electrical and mechanical features of new types of apparatus and supplies" was moved from the parent company and the operating companies to Western Electric's laboratories. AT&T would prescribe the system's technical requirements, but Western Electric would have specific responsibility for the design and standardization of equipment. Western Electric would continue to respond to customers' orders, but it would do so within the confines of centralized, systematic planning and policy.[32]

· IV ·

In 1913 an assembly of top-level officials from AT&T, Western Electric, and Bell operating companies met to discuss the role of Western Electric in the Bell Telephone System. There was no doubt in the minds of these executives that the integration of Western Electric into the system had resulted in a high degree of technological efficiency and considerable financial savings.[33] Yet, the vaunted economies of vertical integration, as they came to be understood by twentieth-century Bell System managers, were only partially anticipated at the time of the acquisition. Of the three fundamental economic motives generally ascribed to backward integration—a fear of supply shortages, a desire to reduce production and administrative costs, and a desire to protect the firm from high prices charged by independent monopolistic suppliers—only the first had been an explicit factor in the recorded thoughts of Bell management in 1882.[34] Adequate productive capacity to support large and rapid growth was, of course, the *sine qua non* of the Bell Company's strategy for gaining control of the field before its primary claims to a patent monopoly expired in 1894.

In addition to the quest for more capacity, the explicit economic motives spurring the acquisition had been: (1) the defensive concern of the Bell Company that Western Electric's ability to generate important inventions might lead to an extremely costly series of patent purchases; and (2) the opportunity for Bell to share, through ownership of its manufacturer, in the profits from the sale of telephone apparatus without having to resort to the loading of royalty fees, a highly unpopular device for extracting revenue from licensees.

Other economic benefits from the acquisition may have been implicitly understood. For example, that Western Electric might achieve lower unit costs per unit production than was possible among smaller, independent businessmen through economies of scale was not discussed by Bell officials before the acquisition, although such economies may have been obvious, nonetheless. Economies of scale derived from divisions of labor, from concentration of capital and management, and from sheer size were well known in the contemporary literature on political economy. In any case, shortly after the acquisition, some economies of scale were noted by Western Electric's board of directors when a committee appointed to consider the consolidation of the Indianapolis and Chicago factories reported its belief that larger-scale production under one management would result in savings for three reasons: "1. more successful and less expensive management; 2. a saving in general expenses of various kinds[;] 3. reduced cost of manufacture due to increase in the product."[35] Some additional economies may have accrued to the Bell Company soon after the acquisition from administrative savings owing to the relative ease of enforcement of contractual obligations and the reduction in correspondence in dealing with one quasi-internal manufacturer rather than several external manufacturers. This would have been intuitively obvious.

Over time, as the operating companies were increasingly brought under the financial control of American Bell, the economic issues involved the measure of savings not so much to Western Electric or to the parent company as to the entire Bell System. By the twentieth century the devolution of a wide range of technical functions upon Western Electric freed the operating companies to specialize more on the task of operating telephone service and AT&T to specialize more on planning and finance for the whole system. The integration and specialization of the organized entities of the Bell System resulted in precisely immeasurable but now widely observed economies in production and distribution. The centralization of production in large volume both reduced the percentage of overhead charges and lowered unit costs. Distribution costs were reduced likewise through the increased volume of direct sales. More accurate centralized forecasts of the equipment needs to the system as a whole reduced the risks of overproduction and the attendant costs in shop wear, inventory maintenance, and insurance. Production costs were also lowered through the introduction of modern principles of "scientific management" in the work place; and the increasingly uniform standardization of hardware helped lower the costs of labor and maintenance in the field through the reduction in the varieties of apparatus that had to be learned and through the smaller number of spare parts that had to be carried.[36]

Equally important amidst the rise of systems engineering in the Bell System were economies of generalized learning resulting from the centralized coordination of design and manufacture and the follow-through into operations. According to Henry B. Thayer, who succeeded Barton as Western Electric's president, the operating companies could now procure all their apparatus from a single source, fashioned "with a single mind," designed "for the use of the whole business." In this way, any one operating company received a direct "benefit of all the experience that we have got in making things for some other place. . . ."[37]

And finally, economies that could be realized from the concentrated purchasing power of a large manufactory were systematically exploited on behalf of the Bell System after the turn of the century. According to Thayer, Western Electric had been sufficiently integrated into the Bell System that its "jobber's natural interest" to charge "a high scale of prices to the customer" had been squelched.[38] Thus, in 1901, when the Philadelphia Bell Company proposed a two-year experiment by which Western Electric was to purchase, warehouse, and distribute all the operating company's supplies, the idea was approved by AT&T. While AT&T president Alexander Cochrane was "doubtful that this plan will fully meet your expectation," he acquiesced in Thayer's reasoning that the centralization of purchasing and distribution would result in

> 1st, The fullest advantage of low prices to be obtained by large quantity in purchases.
> 2nd, A closer touch between the operating departments of the Telephone Companies and the Western Electric Company, enabling the latter company to make a more intelligent forecast of requirements, and therefore a more effective distribution of material.
> 3rd, The saving of hundreds of thousands of dollars per annum in expense of maintaining those departments.
> 4th, The interest on an investment running probably millions in stocks and supplies.[39]

The experiment proved satisfactory, and within twelve years "standard supply contracts" were negotiated between Western and all the Bell operating companies effecting not only the savings that might be expected from Western's enormous bloc purchasing power but also substantial economies in trading expense, in distribution, and in flows of information. On this last point, AT&T controller Charles DuBois stated in 1913 that the supply contracts brought representatives of Western Electric into more direct contact with the telephone companies, especially at the middle management levels, where plant superintendents "had a direct relation to the work requiring supplies." That, declared DuBois, "may have worked to greater advantage than the technical savings that were made through the operation of the contract."[40]

· V ·

By 1915 the Bell Telephone System had come into full flower technologically and organizationally. Technologically, telephony entered the electronic age when Alexander Graham Bell and Thomas Watson held the first transcontinental telephone conversation between New York and San Francisco. The line over which they spoke employed high-vacuum tube repeaters developed by scientists in the Engineering Department of Western Electric.[41] This innovation, which made possible the full-scale technological integration of the nation's telephone exchanges, was paralleled by a wholesale reorganization of the management structure of AT&T, its operating companies, and Western Electric. The process of reorganization, which had begun in AT&T's Long Lines Department in 1908 and then spread throughout the Bell System, clearly divided the plant, traffic, engineering, and commercial functions of the companies, so that information from managers at all levels in each of the system's entities could flow directly to their functional counterparts in the other companies. The reorganization also resulted in a more precise allocation of tasks in the Bell System than ever before, relegating long-distance operations and systemwide financial, legal, business, and technological planning to AT&T; service operations to the operating companies; and research, development, procurement, production, quality control, and central office engineering to Western Electric. The resulting functional organization, modified by the spin-off of Western Electric's Engineering Department to create Bell Laboratories in 1925, endured well into the 1970s.[42]

Also by 1915, when Western Electric was reincorporated in New York, AT&T had boosted its holdings in its manufacturer to 97.4 percent of its common stock.[43] This nearly complete financial takeover was in keeping with Western Electric's having become an inextricable part of the Bell System in functional terms. Considerations that had brought Western Electric under the control of the American Bell Telephone Company in 1882 were important contributing factors, to be sure. Western Electric's capacity and its technological potential proved crucial to Bell's ability to sustain a strong position in the booming and competitive telephone market after 1894. But Western Electric's role in supporting economic development and technological integration of the Bell System had not been so readily foreseen. The subsequent development of large-scale urban exchanges and long-distance telephony generated problems of unexpected complexity. These problems, or constraints on the expansion of the business, weighed heavily on the supply process, forcing a high degree of standardization and efficiency in the manufacture of hardware while compelling its integration with research, design, installation, and maintenance. The technological imperatives driving the business became relatively less a matter of controlling patents and more a matter of integrating the processes of innovation, production, and operations.

Concurrently, the economies derived from increasing scale and consolidation of the production processes were becoming better understood and better exploited. By the turn of the century, integration of Western Electric into the Bell System suggested a role for it as general supplier, or purchasing agent, for the operating companies. In addition to the technological economies associated with the production process, economies could be realized by the operating companies through the "transaction cost savings" derived from the substitution of internal organization for multiple dealings in external markets.[44]

Thus, by the twentieth century many of the technical and business functions originally lodged in the parent company, on the one hand, and in the operating licensees, on the other, had devolved upon Western Electric in a process of integration that far surpassed the American Bell Company's original purposes. In the process, numerous economies of scale and scope were also realized. Bell's arrangement with Western Electric in 1882 to grant it an exclusive market in patented telephone equipment in exchange for control of the manufacturer's capacity and patents was a strategic coup, albeit a coup loaded with far more technological, economic, and organizational significance than anyone imagined at the time.

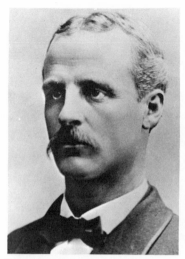

William H. Forbes (1840–97), scion of a Boston investment house, became president of the newly incorporated National Bell Telephone Company in 1879.

Theodore N. Vail (1845–1920) in 1878, when he quit his position as superintendent of the railway telegraph service to join the Bell Telephone Company as its first general manager.

A more mature Vail in 1883.

Charles E. Scribner (1858–1926) went to work for Western Electric as an electrician in 1877. His early work on the development of the multiple switchboard was largely responsible for his company's preeminent position in switchboard manufacture in the early 1880s. During his career, Scribner was granted more than five thousand patents, but none was more important than his early telephonic inventions, including: the jack-knife switch, the "click" busy test, and the bridging line circuit for multiple switchboards. Scribner became chief engineer of Western Electric in 1896, a position he held until his retirement in 1919.

The electrical shop of Charles Williams, Jr., at 109 Court Street in Boston, was the site of all Bell transmitter and receiver production from the early phases of Alexander Graham Bell's experiments until the acquisition of Western Electric. Bell invented the telephone in a room on the top floor.

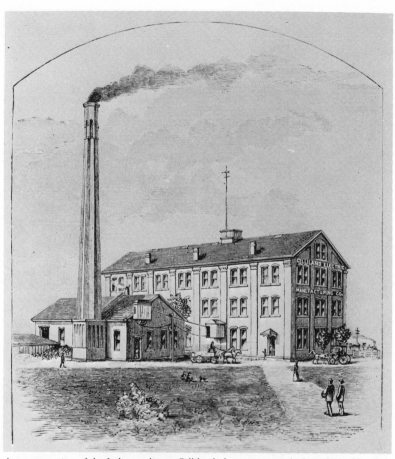

A representation of the Indianapolis, or Gilliland, factory as it might have looked by the early 1880s. One of the four manufacturers recruited by the Bell company in 1879 to manufacture telephone apparatus, the Gilliland shop was known for its advanced production methods. Gilliland was acquired by Western Electric in 1881.

PA-156

A fifty-line Post and Company switchboard of 1878–79 used in Washington, D.C. This represents a complex evolution from the simple five-line board. Pictured are an operator's hand telephone (receiver) and Blake transmitter, a treadle for the generation of power for signaling, and annunciators to announce incoming calls. The board is composed of an upright section and an inclined section, each with horizontal connecting strips in series of four. Vertical, or "line," strips run along the back and underside. Plugs, when not in use, are replaced in a ground strip at the front of the board. At the right is a vertical row of plug holes for trunk-line connections to other boards.

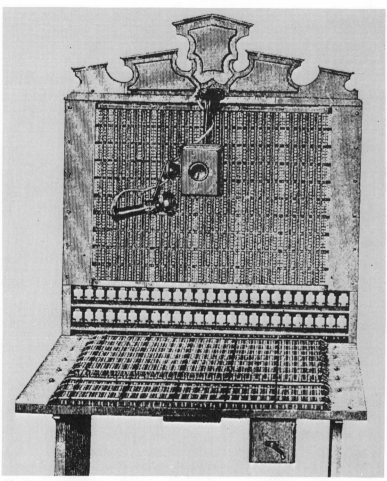

The fifty-line Gilliland switchboard of 1879. Here the annunciators are located between the back and inclined boards, and instead of a treadle, a hand crank is used for signaling. Although this switchboard was of the general plug type, the construction of strips and plugs differed from those of Williams and of Post and Company. Gilliland's construction employed relatively cheap materials, but the board lacked the strength and durability of his competitors'.

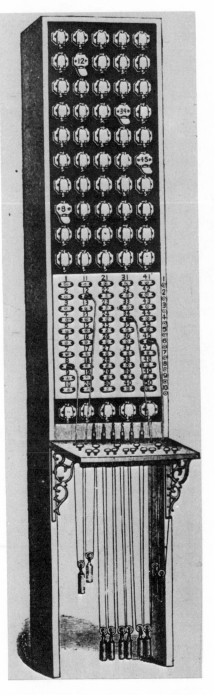

The Western Electric Manufacturing Company's Standard No. 2 switchboard of 1881. This fifty-line board employed cords fitted with weighted pulleys, rather than loose plugs, for connections. It is interesting that Western Electric, the chief manufacturer of telegraph switchboards, was the first to abandon the conventional plug-type design in favor of the more efficient cord. Note also the functional simplicity of the board's design in contrast to the bulky cabinets of the conventional plug boards. The 14-by-60-inch cabinets allowed for better conservation of frontage space, allowing more boards to be placed side by side, in reach of a single operator. In the picture, an operator's telephone and call battery are not shown. The incoming subscribers' lines were connected through spring-jacks to the annunciators (upper portion) and the key pairs, at the front of the shelf. One key of the pair was used to connect the operator's telephone into the calling circuit; the other was used to connect a battery used to signal the subscriber. The cross-section jacks on the right side permitted the use of trunk lines to other boards.

A central exchange office with a capacity of 200 subscribers, Portland, Maine, 1881. What appears to have been a 50-line Post and Company switchboard was supplemented by the addition of 6 25-line boards, connected by trunk lines (not visible) for an exchange of 200 subscribers. Incoming calls on one board requesting subscribers on another board were "trunked through" to a second operator, who was alerted by a special annunciator system (at the left of each board). The work of the operators was coordinated by a central supervisor, who sat at the table in the foreground. The supervisor also handled the transfer of nontelephonic messages coming in and out of the office. The division of labor among operators required carefully organized work plans, which (according to contemporary accounts) varied greatly from one exchange to the next. It is obvious that under such a system, the larger the volume of calls, the more disproportionately complex the problems of coordination.

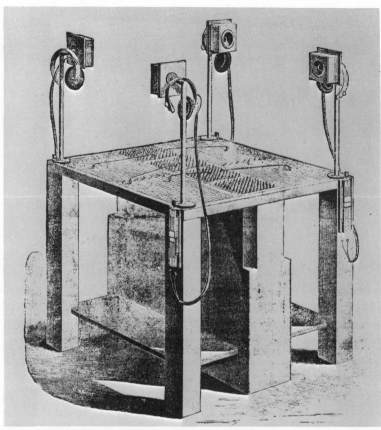

The Law telegraph horizontal switchboard, c. 1880. This system was devised by the Law Telegraph Company in New York to connect law offices and courthouses telegraphically. It was adapted for large-scale telephone use and became the preferred system of large Bell exchanges in St. Louis, Brooklyn, Philadelphia, and the South during the 1880s. Here, a single 34-square-inch board (which could be trunked to other boards) accommodated 400 subscribers and 4 operators. The distinctive feature of the Law System was the use of a separate "call wire" circuit from the subscribers' stations to the exchange. A subscriber initiating a call picked up his transmitter, gave his number, asked a receiving operator for the desired connection by number, and then waited for the connection to be made by a second, switching operator, after which the receiving operator tapped the signal plate. The use of a separate call wire allowed for the elimination of expensive annunciator systems and circuitry, as well as magneto call bells, at the subscriber station. But the system of call wires, each of which connected up to 50 subscribers on a single circuit, required each subscriber to wait until no one else in the system was signaling before placing a call. Operators, moreover, could not speak to the subscribers.

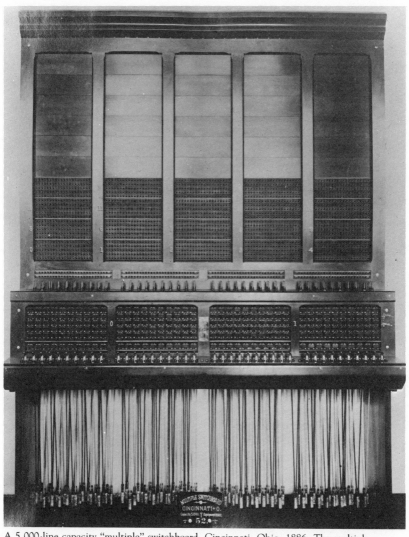

A 5,000-line capacity "multiple" switchboard, Cincinnati, Ohio, 1886. The multiple board, developed by Western Electric after 1879, allowed a single operator to complete the entire call connection from his or her subscribers to any of several thousand subscribers without the use of trunk lines. Here, the subscriber lines for 2,000 subscribers were arranged before two operators, each responsible for receiving 100 circuits through separate answering jacks (the double rows of plug holes just above the cord plugs). By the 1890s single multiple boards were built to accommodate up to 10,000 subscribers. In very large exchanges multiple boards could be trunked to each other.

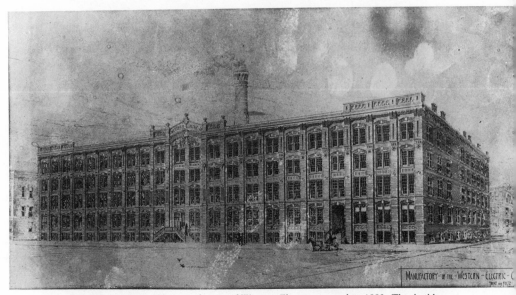

The Clinton Street, Chicago, factory of Western Electric, erected in 1883. This building accommodated more than 1,000 employees by the end of the decade. In 1884 it absorbed the work of the Boston and Indianapolis shops of Charles Williams, Jr., and E. T. Gilliland. By then Western Electric was making telephone and telegraph equipment, electrical lighting apparatus, and burglar and fire alarms. By the late 1890s the Clinton Street facility could scarcely accommodate the explosive growth in Western Electric's telephone and new power apparatus business; in the following decade it was replaced by the Hawthorne Works.

SCIENTIFIC AMERICAN

A WEEKLY JOURNAL OF PRACTICAL INFORMATION, ART, SCIENCE, MECHANICS, CHEMISTRY, AND MANUFACTURES.

Vol. LI.—No. 12.
[NEW SERIES.]

NEW YORK, SEPTEMBER 20, 1884.

$3.20 per Annum.
[POSTAGE PREP'ID.]

ILLUSTRATIONS OF THE AMERICAN BELL TELEPHONE.—[See page 180.]

Research, manufacturing, and operations functions of the American Bell Telephone Company as depicted by *Scientific American* in 1884. The sketch at the top right shows the experimental laboratory at American Bell's Boston headquarters. At right center and bottom, Western Electric craftsmen assemble butterstamp receivers and Blake transmitters. At the top left, Western Electric employees wind and inspect transmitter coils. At left center, a Bell inspector tests a transmitter. The remaining illustrations, from the left center down, show the first experimental telephone, the first switchboard, and the activity of operators at multiple switchboards in the 1884 Boston exchange.

149

Workers at the cotton binding units in a Western Electric facility on Polk Street, Chicago, c. 1890. Just as women replaced men as telephone operators during the 1880s, women were also employed as inexpensive labor for routine tasks in manufacturing. In the 1890s the cable insulation department employed women almost exclusively. Women were also used extensively for equipment inspection.

Blake transmitter assembly craftsmen at the Western Electric shop, Greenwich and Thomas Streets, New York City, c. 1892. By the 1890s, telephones were manufactured in New York and then shipped to Boston for final testing and inspection by the American Bell Company. The individual workbenches and machinery point up the enduring craft nature of telephone equipment manufacture. The patriotic trappings suggest the celebration of some special occasion, perhaps the Fourth of July.

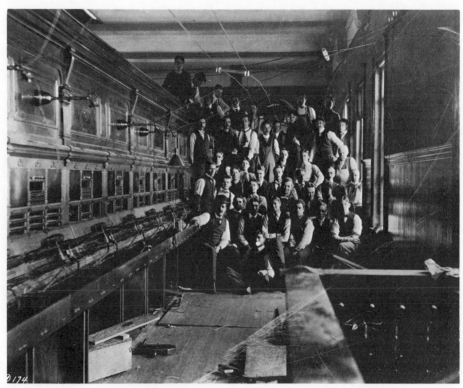

A Western Electric switchboard installation gang in Portland, Oregon, c. 1890. By the twentieth century, Western Electric had assumed responsibility for the installation of large-scale central office equipment in the Bell System. Here a crew of more than fifty men were employed for the installation and wiring of a multiple switchboard complex.

Summary and Conclusion 🌿

THE ORGANIZATIONAL GENIUS OF THE BELL TELEPHONE SYSTEM as it reached maturity in the early twentieth century was its highly integrated vertical structure. Vertical integration supported the centralized planning, production, and orderly implementation of large-scale, capital-intensive, interdependent technological systems at a reasonable cost. Eventually, this would afford efficient telephone service to almost every business, public institution, and household in the United States. The highly integrated vertical structure of the Bell System also supported a remarkably extensive, *decentralized* operating structure well adapted to local demographic, economic, and political circumstances—political circumstances became increasingly important after the resurgence of competition and the increase in the power of state regulatory agencies to oversee telecommunications. This judicious (and, we might add, well-managed) mix of centralized planning and supply functions with decentralized operating functions would be well worth further historical investigation. At each stage of its development, after all, the Bell System, before it achieved its recognizably "modern" form, was evolving in response to newly developing external and internal environments, technologies, and personnel. Nevertheless, what the Bell System became by the early twentieth century in its general outlines, outlines which endured for more than a century, was in large part conditioned by what happened to the Bell enterprise in the first five years of its existence.

In the very early phase of its business, even before Western Union entered the field and before the advent of the commercial exchange, the Bell patentees' lack of capital resources forced them to rely entirely on independent agents for private-line telephone service. Within a short time, the Bell interests created a system of franchises that allowed them to extract a royalty, or license fee, for each telephone placed in service. The proliferation of agencies over a wide geographic area subsequently made Bell's attempt to achieve more control over its business and operating affairs a long and difficult process. In the meantime, Bell owners and managers simply refrained from becoming very involved in the affairs of the local agencies, lacking the administrative resources and finances to do so.

It was a different matter for the manufacture and supply of patented hardware. The early Bell Company's lifeblood was its claim to controlling patents on the telephone and much of its auxiliary apparatus. Financial constraints again prevented the patentees from manufacturing the telephone themselves, but their initial arrange-ment for the production of telephones was as "integrated" a relation-ship as a patent holder could enjoy with an independent producer. Through their strong personal ties and physical proximity to the shop of Charles Williams, Jr., the Bell patentees established an early practice in, and bias toward, keeping the development, production, and distribution of telephone equipment as closely controlled as possible.

These formative tendencies were reinforced during the com-pany's second stage of development, the period of competition with Western Union and the introduction of the telephone exchange. Although its struggle with Western Union impressed Bell with the desirability of taking greater control over its operating agencies as a strategic necessity, financial constraints continued to preclude the firm's seriously contemplating anything more than limited emer-gency acquisitions of local company stock. The central problems for the company in this period—how to meet rising demand under competitive conditions and how to fend off external challenges to its technology—forced Bell to increase the number of its producers and, simultaneously, its sources of innovation. In the process, Bell tried to extend its tight controls over the quality, prices, and distribution of patented apparatus through strongly worded contracts. To some extent, the contracts failed on all three counts, although the expansion of manufacturing capacity at least kept the operating agencies well supplied.

After Western Union's withdrawal from telephony, the Bell Company's perception of its functions broadened considerably. By 1880 the firm was no longer simply concerned with the process by which telephone equipment produced under its patents moved to licensed operating agents. In the pattern followed by many technologically based, capital-intensive businesses, sophisticated financiers and professional managers had taken control of the company's affairs from its inventor-entrepreneur progenitors.[1] Under William Forbes and Theodore Vail the company was becoming more directly concerned with the operations of the agencies, providing financial assistance to many, taking equity in some, and disseminating administrative and technical advice to all.

In addition to taking an active interest in financial development and administrative and technical coordination, Forbes and Vail planned for the long run. Immediate competition may have left the field in 1879, but its specter remained. Anticipating the revival of competition after 1894, they developed strategies for growth and control of the field that encompassed long-term, large-scale development of the market and the aggregation of as much of the patentable technology as they could command. They planned for the penetration of major urban markets, the development of large exchanges, and the thorough interconnection of exchanges by long-distance wires.

By now the financial constraints on the company had relaxed. The firm was increasing its administrative resources and its internal technological expertise. These things boded well for a strong movement forward into the field. But the company's fundamental concerns for patent protection, increasing volume, technological supremacy, and high quality standards, which had guided the firm's policies as a supplier of patented equipment from the beginning, remained important. Thus, before Bell attempted to integrate forward into operations, it sought first to consolidate its control over the production function.

The first major organizational event in the creation of a national telephone network was the acquisition of Western Electric, into which were consolidated the Bell Company's other most important manufacturing licensees. This followed logically from the close control the Bell interests had always tried to exercise over the supply side of their business. A limited form of backward integration was initially contemplated in mid-1880. Then, the Bell Company looked ahead to a time when Charles Williams, Jr., alone would not be likely to satisfy the demand for telephones over the long run. At the same time, problems in enforcing the terms of the 1879 manufacturing contracts for auxiliary apparatus had become all too apparent. These factors, together with the increasing complexity of central exchange office technology, impelled Bell once again to change the manner in which it controlled the production and distribution of equipment.

Reversion to an informal relationship with a single shop based on ties of friendship and proximity—even assuming that one shop could grow enough to meet the Bell licensees' need for rapidly expanding capacity—was no longer possible. The development of telephone equipment, moreover, was no longer the tinkerer's craft that it had been in the early months of 1877, when the first practical instruments were being brought through successive stages of innovation. A more formally structured arrangement that would take into account the need for a large productive capacity and strong inventive resources was needed.

By 1881 a new approach to production was conceived: manufacturing was to be consolidated into one firm with several branches, and the Bell Company was to promote its dominant position in the industry by taking a strong equity position in the manufacturing firm. The sensible (but not necessarily accessible) choice for a partner was Western Electric, with its highly advanced switchboard technology; its strong stable of inventors, engineers, and other skilled machinists; its control of certain transmitter patents; and its preeminent size among electrical manufacturers. Western Electric, of which Western Union owned one third and Anson Stager (a Western Union director and vice president) owned another third, had extensive experience in designing and producing telephone equipment and other electrical supplies. It was only when

Stager fortuitously lost his position at Western Union, for reasons peculiar to that company's history, that Western Electric was delivered into the hands of the American Bell Company.

The reconsolidation of manufacturing on a large scale under Bell Company control set the stage for unanticipated vertical developments, as the dynamic problems of systems engineering replaced the sheer accumulation of patents as the central technological concern of Bell management. By the second decade of the twentieth century—when the Bell System had become a highly integrated, functional organization in which specialized managers and engineers planned, implemented, operated, and maintained a nationwide communications network on a continuous round-the-clock basis— Western Electric had become the linchpin of the enterprise. Having been acquired for its capacity, its patents, and its potential for generating more patents, Western Electric absorbed a wide array of technical functions over time: fundamental research, design and development, maufacturing, quality assurance and control, procurement of outside supplies, and installation of central exchange office equipment. Western Electric, in short, had become AT&T's vehicle for controlling the technical activities and technological development of the Bell System.

· II ·

How does all this relate to the general pattern of vertical integration of large corporate enterprises that emerged in the United States after 1870? In its broad outline, the Bell System's organizational development paralleled that of the dominant national firms in American business before World War I. As with all the firms that pioneered in large-scale, national organization in the late nineteenth century, Bell found that without an underlying strategy of vertical integration, it could not support a viable long-term strategy of extensive horizontal operations.[2] But in the particular purpose and development of its vertically integrated structure the early history of the Bell System stands distinctly apart. Bell's original purpose in integrating forward had less to do with the more common pattern of expanding markets for the distribution of mass-produced goods than with the company's desire to sustain control of its operating companies after the expiration of its patents. This was important not simply because Bell wanted to secure the established outlets for the distribution of its telephones but because the company was already coming to understand that its long-term competitive health rode on its ability to achieve coordination of the operations of telephone *service* at a high, uniform standard. Furthermore, while the dominant pattern during the 1870s and 1880s was one of forward integration, generally involving producers seeking to combine mass distribution with a ready capacity for mass production, for the Bell entrepreneurs the situation was reversed. In their case, the quest for capacity followed the pull of the marketplace, and the market was always threatening to outrun capacity.

Particularly important for Bell in the years prior to the acquisition had been its determination to control its sources of supply for their potential, patentable innovations in telephone technology. Protecting and extending its patent position, no less than assuring adequate capacity, was for Bell a driving impetus toward vertical integration. Quality and price, other common factors of concern in cases of backward integration, were also important to Bell, although in Bell's case the issue of price surfaced as the problem of minimum price maintenance rather than a quest for the lowest charge. Economies of scale through the aggregation and centralization of production were not explicitly sought by Bell in its plans to consolidate and acquire a manufacturer, although they were hinted at in the after-the-fact announcement in the 1882 annual report.

Following the acquisition of Western Electric, the American Bell Company could not maintain (as many of the firms of its day did) merely a passive concern with its source of supply.

The principal reason for this was the transformation of telephony from a business engaged in the production and distribution of hardware into a service providing communications to multiple points over long distances. The institutional and technical problems of providing this service grew more complicated as the emerging network of Bell telephones, wires, and switchboards grew larger. The vast business-technology complex that the Bell enterprise became by the twentieth century required increasingly close, interactive coordination of research, design, development, manufacture, and operations. It was at the locus of manufacture, where the setting of standards and the process of standardization became crucial, that the integration of these activities was most appropriately focused. The more integration was achieved, the more opportunities for economies could be exploited through the standardization of manufacture, purchasing, and operating methods.

The extent to which the various entities of the Bell enterprise became technologically and economically integrated by the early twentieth century was unique in American business.[3] By World War I, AT&T, more than any other company, embodied an emerging American corporate ideal of the standardization of industrial science and technology through the management and engineering of large-scale, complex systems. This was achieved through the organized integration of all the technical and business functions involved in research, production, and the delivery of service. The process of integration had begun a few years after the Bell Patent Association had brought its telephone to market in 1877, and it had begun more out of necessity than as a result of planning. Only three years lapsed before the definite strategies for integration began to unfold in the minds of the early Bell managers, who, when presented with an unusual opportunity to acquire a major supplier, recognized the importance of having a captive source of supply. However limited their understanding may have been of the long-term economic and technological implications of vertical integration, their acquisition of Western Electric laid the foundation for the modern Bell System's highly integrated, multifunctional network. In acquiring a manufacturer and then requiring it to subordinate its interests to those of the agencies providing telephone service, the managers of the Bell Company forged an institutional arrangement that supported the expansive and integrated development of the American telephone industry.

APPENDIX A 🌿

Date	(Periodic) Output by Manufacturer		Telephones in the Hands of Licensed Agents in the United States	
1 July 1877			234	
1 January 1878	5,600		5,200	
1 July 1878			10,000	
1 January 1879	15,101		17,500	
1 July 1879			34,000	
1 February 1880			55,141	
1 March 1880	57,999		90,873[a]	{ 60,873 Bell Company / 30,000 Gold and Stock
20 February 1881	66,763	{ 50,534 domestic / 16,229 for export	132,692	{ 111,807 Bell Company / 20,885 Gold and Stock
20 February 1882[b]	109,349	{ 73,275 domestic / 36,074 for export	189,374[c]	
20 February 1883	(70,453)[d]		249,711[c]	

Sources: S. L. Andrew and E. W. Foss, "Early Telephone Statistics, 1876–1879" (based on Thomas Watson's reports), 1929, AT&T Archives, box 1006; General Manager's Report, 1880, ibid., box 1007; American Bell Telephone Company, *Annual Reports,* 1881–83.

Note: The difference between the number produced and the number leased in a year can be attributed to instruments returned or destroyed, instruments exported, or instruments retained by the Bell Company for other uses.

[a]This marks the first accounting of instruments from former Western Union agencies (Gold and Stock Company) to the Bell licensees after the 11 November 1879 Bell–Western Union agreement.

[b]By 20 February 1882, Western Electric production of Bell Company telephones was well under way. Prior to that time, Charles Williams, Jr., had been the sole manufacturer.

[c]Includes an unspecified number of Gold and Stock instruments.

[d]Estimated number of telephones manufactured *and* purchased for the year. No data can be found for telephones manufactured.

Appendix B ✒

	Territory					
Date	New England	Atlantic Coast	Midwest	South	West	Far West
31 July 1878[a]	11	3	2	0	0	0
31 March 1879[b]	18	7	4	4	1	0
31 December 1879	43	30	61	45	3	0

Source: Record Book: American Bell Telephone Company Agreements, vol. 1, original in AT&T Archives.

Note: This is an account of only those licensees who had formal territorially exclusive contracts with the Bell Company. Many telephones were placed into service under less formal agreements throughout the United States in 1877 and 1878. The first formal agency contract covering a wide territory was issued to the Michigan Telephone and Telegraph Company in the fall of 1877 (see appendix D). It became regular practice after February 1878.

[a]Eve of the contract between the Bell Telephone Company and Charles Williams, Jr.

[b]Soon after the incorporation of the National Bell Telephone Company.

APPENDIX C 🌱

Leasing Telephones

One of the most important business decisions in the history of the Bell System was the first one made by the Patent Association as it placed the telephone on the commercial market in the spring of 1877. According to company tradition, the decision to lease rather than sell telephones was inspired by the experiences of Gardiner Hubbard and Chauncey Smith, attorney for the Patent Association (and former counsel for the Gordon McKay Shoe Machinery Company, which leased its patented machinery in return for a royalty for each pair of shoes produced). Motives for the decision have been related to the greater profitability of leasing a single instrument over a long term and to the enhanced measure of control over the business provided by the leasing of telephones under patents.[1] The leasing of telephones has also been tied to a conception of the telephone business as a technologically integrated service.[2] From the contemporary record, it is possible to glean some of the reasoning behind the leasing policy on all three of the above points: the financial benefits of leasing, the control of patents through leasing, and the service benefits of leasing.

The long-term financial benefits are perhaps self-evident, although they are rarely discussed in the contemporary record. It appears, however, that the patentees regarded leasing as a means by which they could be fairly compensated for the high risks of marketing a novel technology under several resource constraints. As William Forbes, the second Bell Company president, put it in a letter to an investor in 1880:

> Broadly stated, our company had to decide whether it would sell telephones and let them go into [illeg.] all purposes, for what they would bring, before their use and value was once greatly ascertained, a step without recall, or whether we should put them out at an annual rental for various uses as they should be discovered. Keeping the title and charging according to the use.
>
> We adopted the latter, as the only way to reap the return for the heavy outlay of money and work and risk we had to undergo.[3]

Equally compelling was the patentees' recognition that the telephone instrument, though patented, presented few technical barriers to entry. The telephone was easy to copy and manufacture, making patent infringement difficult to monitor and prove under conditions of sale. Chauncey Smith, Jr., remembered his father's understanding "that owing to the comparatively simple nature of the telephone and the ease with which it could be manufactured, there would be in case the instruments were sold, no way of keeping track . . . , while if no instruments were sold and a record was kept of the disposition of each one, the running down and conviction of infringers would be easy."[4]

An early Bell operating licensee, John Ponton, noted in a letter to Alexander Graham Bell in 1898 both the financial and control benefits of leasing. "What struck me most forcibly at that time was the extreme simplicity of your telephone and I kept turning over in my mind the impossibility of making any money on your patent by manufacturing and selling them, which led to the idea of renting them."[5] As for the service benefits of leasing, Forbes noted that under conditions in which customers were being charged for *use*, the company stated "to them what they and we agree to perform." Hubbard advertised at the outset that the lessors would assume responsibility for maintaining the telephone instruments. This commitment was the basis for the gradual assumption by the operating agencies for the provision, maintenance, and replacement of leased equipment once Bell had gained control of its licensees.[6]

The question of selling telephones was often considered and consistently rejected by the Bell Company. In 1880, when the assistant general manager, Oscar Madden, received a request for purchase of a telephone as a museum piece, he replied, "It would be a departure from our custom to sell any of our instruments, even to be retained as you propose, as a relic. Such a door once opened, there is no telling where the thing would end."[7]

APPENDIX D 🐾

The Michigan Telephone and Telegraph Construction Company and the Detroit Electrical Works

An exception to the Bell Company's policy before 1879 to keep production on its patents as close to home as possible exists in the October 1877 agency license granted to the Michigan Telephone and Telegraph Construction Company. The contract,[1] unique in its provision allowing the agency to run for the life of the patents— other early contracts specified no term—also allowed the licensee to furnish and *make* call bells for attachment to telephones leased in its territory. The manufacturing was to be carried out by a subsidiary, the Detroit Electrical Works (which, incidentally, was under contract as an agent for the Western Electric Manufacturing Company). This anomaly in the Bell Company's early manufacturing policy is not easily explained. It is likely that Gardiner Hubbard was not yet so firmly committed to a single licensing policy at the end of 1877, and the Michigan contract may have been a brief foray into an alternative way of dispensing the business. Neither a long-term agency license nor a manufacturing concession would be granted before 1879, and then only under very different circumstances.

In any case, it is difficult to know to what extent call bells were produced in Michigan prior to 1879. Existing correspondence does not provide much help. Slender clues appear in two reports from Oscar Madden and W. A. Jackson to Theodore Vail in 1879 saying, in the first instance, that the Detroit Electrical Works wanted to "go into the manufacture of Telephonic apparatus much more extensively" and, in the second instance, that it wanted a license to manufacture call bells (presumably for sale outside Michigan).[2] A letter from Enos Barton to J. E. Hudson in January 1882 alludes to bells produced by the Detroit Electrical Works, "for the use of their exchanges in Michigan." Barton noted that the Michigan contract was due to expire in March.[3]

Appendix E ❧

Manufacturing Contracts, 1878–1882

The manufacturing contracts struck between the Bell telephone patentees and Charles Williams, Jr. (1878), The Electric Merchandising Company (1879), and The Western Electric Company (1882) provide, in formal, legal terms, some insight into the increasing complexity of the telephone business and the increasing sophistication of its managers. The first contract apparently did little more than codify a set of informal understandings that the Bell interests had had with Charles Williams, Jr., for more than a year. The second contract was identical for the four manufacturers (in addition to Williams) that the National Bell Company recruited in the spring of 1879 to make apparatus. An addendum to the contract clearly limits production to specified types of apparatus, even though the main text leaves open the possibility for the manufacture of telephone instruments.

Like the first two contracts, the third, between the American Bell Telephone Company and Western Electric, is mainly concerned with the protection of patent rights. It is more precise in its wording with respect to the definition of classes of telephone hardware, pricing, and quality. Section 4(e) sets a requirement for dedicated capacity.

The 1878 and 1879 contracts were handwritten in the original; copies can be found in the AT&T Archives. The 1882 contract was printed; a copy is in the AT&T Archives, and the original is held by the corporate secretary.

· Contract with Charles Williams, Jr., 1878 ·

This agreement made this first day of August 1878 by and between the Bell Telephone Company and Charles Williams Jr. both of Boston, Mass.

Witnesseth

That whereas said Company desires to have Telephones and Magneto Call Bells manufactured to supply the market. And whereas said Williams desires to manufacture such Telephones and Magneto Call Bells.

Now therefore said Bell Telephone Company in consideration of the agreement hereinafter contained hereby agrees with said Chas Williams Jr. to purchase all their Telephones of him during the continuance of this agreement and to pay him therefor $1.60 for each Hand Telephone and $2.45 for each Box Telephone—all Telephones to be subject to the inspection and acceptance of the Company's Superintendent. Said Company further agrees with said Williams that all bonds given by the Agents of said Company to said Company shall be held by said Company for the benefit of said Williams and his representatives so far as the same relate to and cover Bell Calls furnished to such Agents or any of them by said Williams under and in accordance with the terms of this contract and that it will protect and save harmless said Williams and his representatives from any and all suits for infringements of any patents upon said Bell Calls.

Said Company further agrees with said Williams that not more than one half of the Bell Calls furnished by him under this agreement shall be leased. Said Company further agrees that if it makes any changes in the Hand or Box Telephones or any of the parts of the same it will indemnify said Williams or his representative for all loss which he or they may sustain by reason of such change of any of said parts already manufactured and on hand to the number of not exceeding five hundred of each part of the Hand Telephones and three hundred of each part of the Box Telephones.

And said Charles Williams Jr in consideration thereof hereby agrees with said Bell Telephone Co that he will manufacture and furnish all Telephones ordered by said Company promptly and subject to the approval of the Company's Superintendent at the price of $1.60 for each Hand Telephone and two dollars and 45/100 $2 45/100 for each Box Telephone.

And that he will manufacture Magneto Bell Calls as ordered by the Company or their Agents all such calls to be subject to inspection and approval of the Company's Superintendent and sell them to all such Agents at eight dollars—and whatever calls said Agents wish to lease not exceeding one half of the whole number furnished, said Williams agrees to furnish to said Agents at an annual rental of Five Dollars except in those cases where the Bell Telephone Company have made contracts to pay a commission on Rentals of Bell Calls in which case the rental shall be Four Dollars.

Said Williams shall be entitled to draw on the consignee, at the rate of three dollars for each bell call shipped, upon shipment thereof. It being understood and agreed that the Company are to be in no way accountable to said Williams for payment for said bell calls except in cases covered by said bonds.

It is further understood and agreed that said Company shall have the right to terminate this agreement at any time upon the failure of said Williams to furnish said Telephone and bell calls promptly and to the satisfaction of said Superintendent and said Agents—and either party may terminate this agreement upon ninety days notice in writing to the other.

If advantage is taken of the right herein provided to terminate this contract without the ninety days notice as above provided material for construction of Telephones in process shall be purchased by Bell Telephone Co at a fair valuation to be mutually agreed upon.

Said Bell Telephone Company are to use their best endeavors to induce their Agents to purchase the Supplies needed in their business from the said Chas Williams Jr.

Said Williams to pack and deliver to depot or shipping place in Boston said Telephones and Bell Calls free of expense to the Bell Telephone Co or their Agents.

Signed the day and year first above written.

[Signed] Thomas Sanders

[Signed] Charles Williams Jr.

[*Signed*] George L. Bradley

[*Signed*] Gardiner G. Hubbard

· Contract with the Electric Merchandising Company, 1879 ·

This agreement made this eleventh day of June 1879, between the National Bell Telephone Company, a corporation organized under the laws of the Commonwealth of Massachusetts and doing business at Boston in said Commonwealth, party of the first part and the Electric Merchandising Company, a corporation organized under the laws of the state of Illinois and doing business at Chicago in the state of Illinois, party of the second part, witnesseth:

Whereas the said party of the first part now owns and controls, and may hereafter own and control letters patent of the United States for improvements in apparatus and instruments relating to telephony and telegraphy and whereas the said party of the second part is desirous of manufacturing apparatus and instruments containing said improvements, now therefore, the said parties have agreed as follows.

First. The said party of the first part hereby licenses and empowers the said party of the second part to manufacture, subject to the conditions hereinafter named, under the said letters patent for improvements in apparatus and instruments relating to telephones and telegraphy now owned and controlled, or which may hereafter be owned and controlled, by said party of the first part such instruments and apparatus as may from time to time be designated by said party of the first part by an agreement in writing additional to this license.

Second. The said party of the second part agrees to pay to the party of the first part on the thirtieth day of June or within thirty days thereafter in each year as a license fee for each apparatus or instrument manufactured by said party of the second part under the letters patent aforesaid such a sum upon each instrument or apparatus as may be fixed from time to time by the party of the first part, to be the same to all manufacturers and not to be more than twenty per cent. of the amount to which said instrument may be sold to agents, excepting however, that in case said party of the first part may have agreed or shall agree to pay a certain sum as a license fee or royalty on each instrument or apparatus manufactured or for each instrument or apparatus for each year of its use then the sum fixed by the party of the first part to be paid as a license fee by said party of the second part shall be regardless of its proportion to the sum for which such instrument may be sold and provided that no license fee shall be paid on any instrument or apparatus except those manufactured after the license fee on said instrument shall have been fixed by the party of the first part.

Third.   The party of the second part agrees not to furnish the said apparatus or instruments to any parties except those authorized by the party of the first part and then only under such restrictions and conditions as the party of the first part may fix from time to time, such restrictions and conditions to apply alike to all agents and manufacturers.

Fourth.   The party of the second part agrees that all instruments or apparatus manufactured under this license shall be in every respect up to and according to a certain standard and shall be in accordance with designs and specifications to be fixed by or acceptable to the said party of the first part and be subject to the inspection of the party of the first part.

Fifth.   The party of the second part agrees that the said instruments or apparatus which are manufactured under this license by the said party of the second part shall be sold to the agents of the party of the first part at a price to be fixed from time to time, but not oftener than once in each six months, by said party of the first part by an agreement in writing additional to this, after full consideration of the cost of manufacture and to be the same price for all agents and manufacturers and to be the cost of manufacture as herein provided together with such license fee as may be fixed by the party of the first part and a fair profit for the manufacturer provided that the price fixed from time to time shall only apply to instruments or apparatus manufactured after the date of the agreement by which said price is fixed.

Sixth.   The party of the second part agrees to make full and true returns to the party of the first part under oath upon the thirtieth day of June in each year during the continuance of this license of all instruments or apparatus manufactured by the said party of the second part under the letters patent aforesaid.

Seventh. The party of the second part agrees to make a statement in writing to the party of the first part once in each week covering the week ending on Saturday night of the number of said instruments or apparatus of each kind ordered, by whom ordered and the dates on which such orders were received, the number of orders filled, to whom filled, with the dates when filled and also the number of said apparatus or instruments manufactured during the week and to mail said statement to the party of the first part not later than Tuesday in the next week.

Eighth. The party of the second part agrees that each instrument or apparatus manufactured under this license shall be numbered consecutively with such series of numbers as may be indicated by the said party of the first part and that on each instrument or apparatus, in addition to the patent marks and numbers required by law, shall be plainly lettered the following:

> Made for the National Bell Telephone Company and after that the name and place of business of the party of the second part.

Ninth. The party of the second part agrees that in case the party of the second part or any of its agents, employees or servants shall make any new and useful improvement upon any of the above described instruments or apparatus, the exclusive use of such improvements shall be conveyed to the party of the first part subject to a fair price to be agreed upon, or in case a price cannot be agreed upon then the price to be determined by three referees, one to be chosen by the party of the first part, another by the party owning or controlling the invention and the third by the two so chosen.

Tenth. Upon the failure of the party of the second part to comply with the above conditions of this license, the party of the first part may terminate the same by serving a written notice upon the party of the second part but the party of the second part shall not thereby be discharged from any liability to the party of the first part for any license fees due at the time of the service of this notice.

This agreement shall continue for five (5) years unless sooner terminated either by failure of the party of the second part to comply with all the conditions of this agreement or by six months' notice from either party in writing of a desire to terminate. Whereupon at the expiration of said six months it shall be terminated.

In witness whereof the parties above named have executed this agreement by the proper officers duly authorized so to do on the day and year first above written.

The National Bell Telephone Co.
By Its General Manager
[*Signed*] Theo Vail
Electric Merchandising Co.
[*Signed*] By Geo H. Bliss, Prest

Approved:
[*Signed*] Geo L. Bradley
Vice Pres

In accordance with an agreement entered into this day by and between The National Bell Telephone Company, party of the first part, and the Electric Merchandising Company of Chicago, party of the second part it is hereby agreed as follows

That the party of the second part shall be and is hereby authorized to manufacture instruments commonly known as
Magneto Call Bells
Hook District Bells
using in combination with such instruments the switch commonly known as the
Secrecy Switch, and Automatic Switch,
also Annunciator Drops.
The prices agreed upon as follows:
Magneto Call Bells with or without secrecy and automatic switches and battery cutoff Eight Dollars, $8.
Hook District Bells with secrecy switch, automatic switch and battery cutoff, Three 25/100 Dollars, $3.25.
Annunciator Drops, One 50/100 Dollars, $1.50
Until further agreement no royalty or license fee shall be required to be paid the party of the second part.

In witness whereof the parties to these presents have hereunto set their hands and seals this eleventh day of June A.D. 1879.

> The National Bell Telephone Co.
> By Its General Manager
> [*Signed*] Theo N. Vail
> Electric Merchandising Co.
> [*Signed*] By Geo. H. Bliss Prest

Approved:
[*Signed*] Geo. L. Bradley
Vice President

· Bell–Western Electric Manufacturing Contract, 1882 ·

THIS AGREEMENT made this Sixth day of February A.D. 1882, by and between THE AMERICAN BELL TELEPHONE COMPANY, a corporation created and existing under the laws of the Commonwealth of Massachusetts, of the first part, and THE WESTERN ELECTRIC COMPANY, a corporation created and existing under the laws of the State of Illinois, of the second part,

WITNESSETH:

WHEREAS said second party desires to be employed by said first party to manufacture the telephones required by it, said first party, and to be licensed to manufacture telephonic appliances and apparatus other than telephones under Letters-Patent now owned or controlled or which may hereafter be owned or controlled by said first party, and WHEREAS said second party owns certain inventions, patents, and rights under patents, relating to telephones and telephonic appliances,

NOW THEREFORE IT IS AGREED:

1. Said second party agrees that it will, if said first party shall elect to purchase the same at any time within six months from the date hereof, sell, assign, and transfer to the first party at the actual cost thereof to it said second party, either, or any, or all the inventions in electrical speaking telephones, or improvements therein, or applicable thereto, which it said second party now owns, and

further that it said second party will from time to time, if and whenever it shall acquire any inventions in electric speaking telephones, or improvements therein, or applicable thereto, forthwith notify said first party thereof and that it will, in case said first party shall elect to purchase the same within six months after such notice, sell, assign, and transfer the same to said first party at the actual cost thereof to said second party.

The word "telephone" as used in this contract includes all instruments employed for the electrical transmission of articulate speech, including therein all attachments and devices which serve to cause, to improve, or to modify the articulating current or the effects thereof.

2.   Said first party agrees that it will employ said second party during the existence of this contract to manufacture telephones required by it, The American Bell Telephone Company, whether for use in the United States, or for Export, upon the term and conditions following, that is to say:

Said telephones shall be manufactured only at such of the manufactories of said second party as said first party may from time to time designate.

At all times during their manufacture and upon their completion the instruments and the materials employed shall be subject to the inspection and acceptance of Superintendents and inspectors appointed by said first party.

The price to be paid therefor shall be the actual cost thereof with an addition of twenty per cent of such cost as manufacturer's profit.

Said second party shall promptly manufacture and supply all telephones ordered by said American Bell Telephone Company, and shall provide itself with facilities for making them in such quantities as the course of business may demand.

Said telephones shall be of such patterns and styles and shall embody such inventions and improvements as said first party may from time to time prescribe and shall be made in all respects as they shall direct as to style, material, workmanship, and finish.

In addition to the patent marks and figures said telephones shall bear such other marks and figures as said American Bell Telephone Company shall from time to time direct and no others.

Said manufacture shall be subject to such regulations as may in the opinion of the first party be necessary for its protection and as it shall from time to time prescribe to said second party.

Except as in this article provided for the second party shall not engage in the manufacture of telephones within the United States.

Said second party faithfully complying with the terms of this employment, said first party agrees that it will during the existence of this contract employ no others to manufacture telephones.

3.   Said second party hath granted and doth hereby grant to the first party the sole and exclusive right and license during the full term for which patents thereon have been or may be granted to make, use, and sell, and to license others, except only the existing licensed manufacturers of said first party named in Article 4 hereof to make use, and sell in call bells, switches and other telephonic appliances, any inventions or improvements therein, which it said second party does now or may hereafter own or control, in whole or in part, by contract or otherwise, whether patented or not, "Telephonic Appliances" include calls, switches, switch boards, annunciators, exchange furniture, and other apparatus and devices adapted for use on or for telephone lines, except telephones as above defined.

The second party agrees that it will from time to time and at all times do, execute, and deliver such other and further acts and instruments, if any, as may be necessary to secure to and vest in said first party such sole and exclusive right and license.

4.   Said first party hereby grants and agrees to grant (subject, however, as to inventions hereafter directly acquired by said first party to the provisions of clause 3 of this Article) to the second party a license exclusive except as hereinafter stated, upon the terms and conditions and subject to the limitations herein expressed during the full term for which Letters-Patent thereon have been or may be granted to make and to sell call bells, switches and other telephonic appliances as above defined embodying any inventions or improvements thereon which it said first party does now or may hereafter own or control (including herein the inventions and improvements licensed to it, said first party, by said second party in the previous article hereof); but said second party shall and said second party hereby agrees that it will,

(a) Assume and pay or discharge any and all royalties and other obligations which said first party is or may be bound to pay or perform on or on account of said inventions or any of them.

(b) Within the United States sell such apparatus only to the Licensees of telephones of The American Bell Telephone Company and to them at prices not exorbitant or unreasonable, and under the terms and limitations hereinafter stated, such prices are to be uniform for the different classes of Licensees of said American Bell Telephone Company.

(c) In selling such appliances for use in foreign countries it will in each case retain the title thereto until they are landed in the foreign country to which they are sent; and will insert in its bill heads therefor the following, or such similar notice as the first party may require viz.

"The articles herein named are to remain the property of the Western Electric Company until landed in _____. It is understood and agreed to by the purchaser that no right to use the articles herein named in the United States, and no right to use any patents whatever granted by the United States is conveyed hereby, no consideration having been paid for such use," and will insert in its bill heads for appliances which it is authorized to sell to Licensees of The American Bell Telephone Company in the United States the following or such similar notices as the first party may require, viz:

"The call bells and other telephonic appliances, included in this bill are made under patents which belong to the American Bell Telephone Company, or which they have the exclusive right to use, and are only licensed to be used in connection with telephones licensed by said company and at Stations using such telephones; and the purchaser by accepting them agrees not to use them otherwise nor to dispose of them to anyone except those now licensed."

And will conform to such other similar regulation as said first party may reasonably require for its protection and may from time to time prescribe to said second party.

(d) Mark such apparatus in addition to the patent numbers and marks with such other numbers, marks, notices, or names, and no others, and in such manner as said first party shall from time to time prescribe.

(e) Promptly manufacture and supply all such apparatus as it is hereby authorized to supply, and of such patterns and styles and embodying such inventions and improvements as may be ordered by the parties to whom it is authorized to dispose of the same and will provide itself with facilities for making such apparatus in such quantities as the course of business may demand.

(f) Will on the first days of January and July in each year, during the continuance hereof, making full and true returns under oath of all and singular the telephonic appliances and apparatus manufactured hereunder.

(g) And except as in this article provided shall not engage in the manufacture of telephonic appliances within the United States.

(h) The party of the second part agrees that all instruments or apparatus manufactured under this license shall be of the most approved forms and first class in point of material, finish and workmanship and shall be subject to the inspection of the party of the first part.

(i) The license hereby granted is not exclusive as to inventions or improvements in cables, conductors, insulators, or the manner of constructing and arranging lines (apart from switches and instruments) and

The rights in this article granted are subject to the existing rights of Post & Company of Cincinnati, Ohio under contract dated June 27, 1879 with the National Bell Telephone Company; of Davis and Watts of Baltimore, Maryland under contract dated June 24, 1879 with the National Bell Telephone Company; of the Electric Merchandising Company under contract dated June 11, 1879 with the National Bell Telephone Company; and to the manufacturing rights of the Telephone and Telegraph Construction Company under contract dated October 24, 1877 with Gardiner G. Hubbard, Trustee. But the first party agrees that it will within six months after request therefor from said second party give notice to terminate such of said last named contracts as can be so terminated as therein provided; and it is agreed that the second party shall until such termination be entitled to receive all royalties accruing to the first party thereunder after July 1, 1881, on account of manufactures less only royalties which said first party may be bound to pay on account thereof.

(j) In case said first party shall hereafter acquire any invention or inventions in telephone appliances as above defined, other than such as it may hereafter acquire by license from the second party under article 3 hereof, it shall forthwith notify said second party thereof, and such invention or inventions shall in case said second party shall so elect within six months from the date of such notice, come under the provisions of this Article 4 hereof with like effect as if such invention or inventions were now owned by said first party; but the second party, in case it shall elect to have such license, shall forthwith upon making such election repay, to said first party the cost of such invention or inventions and assume any and all royalties and other obligations which said first party may be bound to pay or perform on or on account of such invention or inventions.

5.   WHEREAS the one prosecution of the business of said Telephone Company and of its Licensees, users of its telephones; requires that telephones and telephonic appliances of the most approved forms and workmanship be promptly furnished and as herein provided, and WHEREAS the said Telephone Company is unwilling to permit the reasonable expectations of its licensees in this regard to be disappointed and whereas the time required to ascertain judicially whether the second party shall or shall not have failed to perform its obligations hereunder would cause a delay which might work great irreparable and unascertainable damage to said telephone company and its said licensees, now it is an integral part of this contract and is also a limitation of and an exception to the license hereby granted and the employment hereby contracted for that:

(1) Whenever the Directors or Executive Committee of said telephone company shall be of opinion that the second party has failed in its obligations in this respect, or that there is imminent danger or probability that it will so fail, the said telephone company may without notice or demand, forthwith manufacture or cause to be manufactured elsewhere, either by itself or others, such instruments as in its opinion may be required to meet the emergency, and may continue and prepare to continue such manufacture elsewhere so long as in its opinion may be needful; and for the purpose of obtaining such supply, the opinion of its Directors of its Executive Committee arrived at in good faith, shall be conclusive as to its right so to obtain a supply elsewhere, and its action based thereon shall not

be restrained by any court or judicial power whatever; but such decision shall not terminate this contract and license nor shall it prevent the second party from furnishing apparatus to any persons who may lawfully order the same according to the terms hereof.

(2) If it shall thereafter be determined by agreement of the parties or by the final judgment of any Court of competent jurisdiction that the opinion so acted on by said telephone company was erroneous, then the consequences of such error shall be that said telephone company shall pay to said second party all the profits which it, said second party, would actually have realized and received if such supply so obtained elsewhere had been obtained of said second party and said decision of said telephone company shall not authorize it thereafter to obtain such supply elsewhere.

(3) If it shall be determined by agreement of the parties or by the final judgment of any Court of competent jurisdiction that said second party did violate or fail to comply with or has violated or failed to comply with any of the terms of this agreement then this license to manufacture telephonic appliances shall be deemed to have been non-exclusive and the obligation to employ the second party to manufacture telephones to have ceased as from the time of such violation or default and said license shall remain non-exclusive and the first party shall be under no obligation to employ the second party to manufacture telephones unless and until said second party shall upon notice from said first party have remedied or repaired such default, violation or neglect and shall have repaid and made good to said first party and all loss, cost, damage and expense occasioned thereby or resulting therefrom, and shall reasonably satisfy said first party that it is prepared and intends to conform hereto, whereupon said license shall again become exclusive and the obligations to employ the second party to manufacture telephones shall revive, such license and such employment being however subject to all the terms, conditions, limitations, and stipulations, inclusive of this article, herein contained.

6. Existing interferences in cases of inventions in telephonic appliances shall be disposed of as counsel of the parties hereto may advise, to recognize and protect the rights of the several inventors to their respective inventions subject to the decisions of the Patent Office.

7. The first party agrees that it will, so long as and in so far as the license herein granted the second party is and shall remain exclusive, consent to the use of its name either alone or together with that of the second party as the case may require, in all such suits as counsel of the parties may advise are necessary or proper for the protection of such rights against infringement, but the whole expense of such litigation shall be borne by the second party.

8. This contract shall remain in force until it shall be determined by the mutual agreement of the parties hereto.

IN WITNESS WHEREOF The American Bell Telephone Company has caused these presents to be signed in its name and behalf by William H. Forbes its President and its corporate seal to be hereto affixed and the Western Electric Company has caused these presents to be signed in its name and behalf by Anson Stager, its President and its corporate seal to be hereto affixed the day and year first above written.

The American Bell Telephone Company
[Signed] by W. H. Forbes, President
Western Electric Company,
[Signed] by Anson Stager, President

Attest,
[Signed] S. G. Lynch, Secretary

Essay on Sources ❧

· Archival Sources ·

The American Telephone and Telegraph Company has an extraordinary documentary record of the first fifty years of its history. In its archives in New York City, AT&T has stored a rich collection of business records dealing with various elements of policy, strategy, finance, and operations from the late 1870s to the 1920s. The collection appears to have been assembled and organized into a working archives in the 1920s by William Chauncey Langdon, whose work on the records has the earmarks of a highly personal labor. With Langdon's passing, no means for the ongoing accession of documents was put in place, and so the collection grew very little. The archives remained an active source of information about the company, however. The company's records were thoroughly mined by the Federal Communications Commission (FCC) in the 1930s, and they have been kept by the company, for internal purposes, to the present day. Recently, the reorganized and revitalized archives has mounted an effort to bring the holdings up to date in accordance with modern archival standards.

For the researcher, the catalogue provides titles of document folders. These titles, the product of a catalogue revision in the 1950s and 1960s, do not correspond to citations used by older works on company history and reflect an arrangement that divorced many of the records from their provenance. For this latter reason and because the folder titles are the only describers, the researchers must do some imaginative cross-referring in order to cover particular subjects. AT&T personnel who have worked with the collection know both its contents and idiosyncracies well and are good guides to specific categories of information. Most of the documents have been photographed, and for reasons of security and preservation, it is to microfilm or microfiche that the researcher is ordinarily allowed access.

The holdings are divided into three main categories: (1) boxed correspondence (both incoming and outgoing), reports (formal and informal), memoranda, memoirs, and miscellaneous items (including some secondary sources); (2) voluminous runs of presidents' and general managers' letterpress books for outgoing correspondence pertaining to both policy and operational concerns; and (3) rare books and typescripts regarding technical and legal matters. Particularly useful in this last category are bound collections of testimonies and evidence relating to some of the early patent suits, printed proceedings of the National Telephone Exchange Association (a convention of Bell licensees that met during the 1880s), and printed and typescript proceedings of internal Bell System switchboard and cable conferences which were held occasionally from 1887 to 1895.

The AT&T Corporate Archives is the best and in some cases the only accessible source for biographical materials on some of the significant actors in early telephone history. This is especially true for men such as Hubbard, Sanders, Watson, and Williams. An important exception, of course, is Alexander Graham Bell, whose life has been amply chronicled and whose papers are located at the Library of Congress (having been transferred there in 1975 from the National Geographic Society).

In contrast to AT&T, Western Electric preserved relatively little of its early historical record. There is, however, a small collection of Western Electric records now held by the AT&T Bell Laboratories Archives, Murray Hill, N.J., containing some original correspondence relating to the company's pre-Bell history (1869–82). The Western Electric collection also contains some biographical sketches of some of the company's more illustrious personnel, a complete run of the *Western Electric News* (1912–32) (on which I have relied for memoirs of some of the early managers and electricians), and annual reports dating from 1904.

A significant collection of historical records relating to research and development is being assembled into a modern archival facility at the AT&T Bell Laboratories Archives in New Jersey. The

Bell Laboratories R&D Archives contains files of prominent laboratory executives, laboratory notebooks, and a large quantity of American Bell Company correspondence running continually from 1883 to 1900. Because Western Electric became the principal research arm of the Bell System from 1907 until the founding of the Bell Laboratories in 1925, there are important materials relating to the manufacturer's history for that period in the R&D collection in New Jersey.

As for Western Union (which figures prominently in this study as the early rival to the Bell enterprise), little in the way of primary records exists outside the AT&T Archives. At Western Union's Corporate Information Center, in Upper Saddle River, New Jersey, there is a collection of nineteenth-century annual reports and contemporary trade journals supplemented by some internally written commemorative histories.

· Published Works and Other Public Sources ·

· *General Histories of the Bell System and the U.S. Telephone Industry*

The first great wave of general historical interest in the telephone industry came with the launching of the Federal Communications Commission's investigation into the corporate and financial history of the Bell Telephone System in 1935. After four years of intensive research in the records of AT&T and many of its subsidiary companies, the commission published its *Investigation of the Telephone Industry in the United States* (Washington, D.C., 1939), which was supported by seventy-seven volumes of associated staff exhibits. Many of the exhibits treat, in varying degrees of detail, the early history of the Bell enterprise with respect to its corporate and capital structures, its technological development, its methods of protecting its patent monopoly, and its rates. The exhibits refer to important primary sources, but their texts vary greatly in their sophistication and scholarly objectivity. In sum, the exhibits, the *Investigation,* and a preliminary publication entitled *Proposed Report on the Telephone Investigation* (Washington, D.C., 1938) reflect a clear bias against

AT&T's status as a large, monopolistic business corporation. AT&T issued a report citing a number of errors in the commission's work entitled *Brief of Bell System Companies on Commissioner Walker's Proposed Report on the Telephone Investigation* (New York, 1939). An offshoot of the commission's work was N. R. Danielian's *AT&T: The Story of Industrial Conquest* (New York, 1939), a scholarly synthesis of Bell System history integrating technological, institutional, and economic aspects of the history. Its ideological position reflects that of the commission's work generally, but Danielian's analysis is powerfully argued and intelligently wrought. Collectively, the commission's reports and Danielian's history remain important points of departure for the serious student of the early history of telephony.

A second and larger wave of historical interest in the Bell System is only beginning to swell in the wake of the centennial of the telephone and the more recent reorganization of AT&T's corporate structure—including the government's forced divestiture of much of AT&T's traditional telephone business. Two centennial works of note are John Brooks, *Telephone: The First Hundred Years* (New York, 1975, 1976), a competent general narrative history written from a boardroom perspective; and Ithiel de Sola Pool, ed., *The Social Impact of the Telephone* (Cambridge, Mass., 1977), an anthology of essays ranging from the scholarly to the promotional on the ways in which the telephone has transformed the culture. Very recent attempts at analysis of the history of both the telegraph and telephone businesses are contained in Gerald W. Brock, *The Telecommunications Industry: The Dynamics of Market Structure* (Cambridge, Mass., 1981), and Paul R. Lawrence and Davis Dyer, *Renewing American Industry* (New York, 1983). In both cases the histories are overtly keyed to the contemporary social concerns of the authors. Brock (like Danielian, an economist engaged in critique) treats the history of AT&T as a problem in a monopolist's attempts to sustain barriers to entry over time in a changing market. Lawrence and Dyer (an organizational analyst and a historian, respectively) are concerned with the changing relationships between institutional arrangements and managerial styles, on the one hand, and market and regulatory influences, on the other.

A concentrated history of the early Bell Telephone business is R. J. Tosiello, "The Birth and Early Years of the Bell Telephone

System, 1876–1880" (Ph.D. diss., Boston University, 1971). Unfortunately, this study has no interpretive framework, but Tosiello's densely detailed narrative serves as an excellent guide to sources in the AT&T Archives. Another thesis, P. Ronald Tarullo, "American Telephone and Telegraph Company: A Survey of Its Development through Basic Strategy and Structure" (Ph.D. diss., University of Pittsburgh, 1976), argues the dubious proposition that AT&T's historical behavior has been unaffected by regulation. Tarullo's treatment of the nineteenth-century history (for which his proposition is somewhat stronger than for the main period of his study) is based on a superficial treatment of skimpy sources.

Some works of limited value, though they are often cited, are: Herbert N. Casson, *The History of the Telephone* (Chicago, 1910); Horace Coon, *American T&T: The Story of a Great Monopoly* (New York, 1939); and Joseph Goulden, *Monopoly* (New York, 1968). An AT&T in-house publication, *Events in Telephone History* (New York, 1983), provides a useful chronology.

· Histories of Telephone Technology

For those who believe that technology is learned best when studied through its morphology, there is good news with respect to telephony. The critical importance of the company's patent base compelled the Bell Company to track and record the evolution of its hardware and systems from almost the beginning. Detailed accounts of the development of even the most mundane pieces of hardware can be found in the AT&T Archives, in legal testimonies, and in contemporary trade journals. Periodically, published works appeared that synthesized and summarized the state of the art. The first of such publications (and still one of the most readable) was Western Union electrician George Bartlett Prescott's *Bell's Electric Speaking Telephone* (1884), 2d ed., rev., *The Electric Telephone* (New York, 1890).

The subsequent literature is vast, and much of its can be traced through the bibliography of M. D. Fagan, ed., *A History of Engineering and Science in the Bell System* (n.p., 1975), which is itself an extensive, encyclopedic source on the development of telephone hardware and systems by category (station apparatus, transmission facilities, switching and signaling apparatus, etc.). For discussions of early telephone hardware some good works are: Kempster B. Miller,

American Telephone Practice, 4th ed. (New York, [1899] 1905); Fred DeLand, "Notes on the Development of Telephone Service," *Popular Science Monthly* 70 (May 1907), 403–12; John E. Kingsbury, *The Telephone and Telephone Exchanges* (London, 1915); Frank B. Jewett, "The Telephone Switchboard—Fifty Years of History," *Bell Telephone Quarterly* 7 (July 1928), 156 ff.; and Frederick Leland Rhodes, *Beginnings of Telephony* (New York, 1929). A very recent attempt to relate the development of telephone hardware and systems to the development of organizational structure is the elegant, written testimony of James Porter Baughman in *U.S. v. AT&T* (1982). In a similar vein is a working paper by Alan Gardner, "Summary of Research on the History of Long Lines, 1894–1927," 21 January 1981, copies of which are on file in the AT&T Archives.

In the past few years, a number of published works have begun to approach the technology of telephony less as a problem in the development of techniques and more as a function of wider business and social phenomena. Two articles by David A. Hounshell examine the invention of the telephone as a way of commenting on aspects of the processes of invention: "Elisha Gray and the Telephone: The Disadvantages of Being an Expert," *Technology and Culture* 16 (April 1975), 133–62; and "Bell and Gray: Contrasts in Style, Politics, and Etiquette," *Proceedings of the IEEE* 64 (September 1976), 1305–14. Sidney H. Aronson, "Bell's Electrical Toy: What's the Use? The Sociology of Early Telephone Usage," in Pool, ed., *Social Impact of the Telephone*, deals intelligently with the problem of public acceptance of a novel technology. George David Smith, "The Bell–Western Union Patent Agreement of 1879: A Study in Corporate Imagination," *American Historical Association Proceedings, 1982* (Ann Arbor, Mich.: University Microfilms, 1982), attempts to relate perceptions of technological innovation to business strategy. In a more popular vein see Michael F. Wolff, "The Marriage That Almost Was," *IEEE Spectrum* (February 1976), 41–51. An unusual but appropriate attempt to see the strategic history of technology as a spatial problem is John V. Langdale, "The Growth of Long-Distance Telephony in the Bell System: 1875–1907," *Journal of Historical Geography* 4 (1978), 145–59.

Articles tracing the institutionalization of research and development at American Bell and AT&T are: Leonard S. Reich,

"Research, Patents, and the Struggle to Control Radio: Big Business and the Uses of Industrial Research," *Business History Review* 51 (Summer 1977), 230–35; idem, "Industrial Research and the Pursuit of Corporate Security: The Early Years of Bell Labs," ibid., 54 (Winter 1980), 504–29; and Lillian Hoddeson, "The Emergence of Basic Research in the Bell Telephone System, 1875–1915," *Technology and Culture* 21 (1981), 512–44. Reich's work, an extension of his Ph.D. dissertation, "Radio Electronics and the Development of Industrial Research in the Bell System (Johns Hopkins University, 1977), is concerned mainly with the strategic implications of industrial research and development, while Hoddeson examines the internal technological and scientific influences on the organization and conduct of research in the corporation. Useful also are Danielian, *AT&T*; Thomas Shaw, "The Conquest of Distance by Wire Telephony," in *The Telephone: An Historical Anthology,* ed. George Shiers (New York, 1977); Neil Wasserman, *The Path from Invention to Innovation: The Case of Long-Distance Telephone Transmission at the Turn of the Century* (Baltimore, 1985); James E. Brittain, "The Introduction of the Loading Coil: George A. Campbell and Michael I. Pupin," *Technology and Culture* 11 (January 1970), 36–57. Together these works represent several unwoven strands of interpretation regarding Bell System research and development. Taken separately, the strands relate to such varied concerns as business and patent strategies, technological bottlenecks, the professionalization of engineering, and the quest for more basic scientific knowledge as those concerns influenced organizational development.

Financial Histories

The FCC's *Investigation* (and its attendant exhibits) was, above all, an attempt to understand the financial development of the Bell System, and it still serves as the best source for those aspects of the company history concerned with the sources and uses of funds and methods of accounting. Because standard procedures for accounting were not developed by the American Bell Company until the mid-1880s, understanding of the financial history for the first several years of the business's development is necessarily more impression-

istic than systematic. Useful summaries of the early financial arrangements supporting the development of the parent corporation
and its evolving relationship to its licensees are J. Warren Stehman,
*The Financial History of the American Telephone and Telegraph Com-
pany* (Boston, 1925); and [William Chauncey Langdon,] *The Early
Corporate Development of the Telephone* (New York, 1935). Although
Stehman's work is lengthy and scholarly, it is somewhat quaint by
modern standards of financial reporting.

A pair of useful specialized studies are *Depreciation: History
and Concepts in the Bell System* (1957), and Laurence E. Steadman,
"Telephone Accounting, 1877–1912: From Early Bookkeeping to
Modern Management Systems" (1981). The former is an in-house
publication; the latter, which deals with the transformation of
systems for gathering information into systems of control, is on file at
AT&T.

· *Biographies and Memoirs*

Published biographies of the founders of telephony are very much in
the heroic mode. Only one, Robert V. Bruce, *Bell: Alexander
Graham Bell and the Conquest of Solitude* (Boston, 1973), is a significant work of scholarship. Bruce's book supplants Catherine
MacKenzie, *Alexander Graham Bell* (Boston, 1928), and a host of
lesser works. Because Alexander Graham Bell was not an active
participant in the development of the corporations created to turn
his invention into a marketable business, his biography is somewhat
peripheral to the business history of the telephone.

For Theodore Vail and William Forbes one can turn to
Albert Bigelow Paine, *Theodore N. Vail: A Biography* (New York,
1921), and Arthur S. Pier, *Forbes: Telephone Pioneer* (New York,
1953). These uncritical and unannotated volumes provide some
useful background but must be treated with discriminating care:
some of the anecdotal content has the ring of apocrypha. A treatment of the same ilk exists for the inventor Emile Berliner: Frederic
William Wile, *Emile Berliner: Maker of the Microphone* (Indianapolis,
1926). A better biography illuminating some of the early engineering history is Frederick Leland Rhodes, *John J. Carty: An Appreciation*
(New York, 1932). Other biographical accounts of the principals of

this history can be traced through unpublished accounts in the AT&T and Western Electric archives and in various Bell System publications. Brief but useful accounts of Gardiner Hubbard, Anson Stager, J. J. Storrow, and Vail are in the *Dictionary of American Biography.*

Thomas A. Watson's memoirs have an authentic ring and offer some rare commentary on the human experience of the telephone enterprise. See his *Exploring Life* (New York, 1926) and *The Birth and Babyhood of the Telephone* (New York, 1913). Aside from printed memoirs in the *Western Electric News* (previously noted), there are important testaments on the early development of telephone production and technology by Vail, Enos Barton, and Henry Thayer in *Western Union v. American Bell Telephone Co.*, exhibit C, "Evidence for the Defendant," 2 vols. (Boston, 1909).

· *Contexts and Background*

The indispensable introduction to the history of large American corporations is Alfred D. Chandler, Jr., *The Visible Hand: The Managerial Revolution in American Business* (Cambridge, Mass., 1977), which examines the rise of big business as the organizational response to the creation of a national market amidst the technological revolution of the latter part of the nineteenth century. *Visible Hand* builds on Chandler's seminal analysis *Strategy and Structure: Chapters in the History of the American Industrial Enterprise* (Cambridge, Mass., 1962), a preliminary attempt to set the American experience in an international context. Chandler, oddly enough, pays little attention to the telephone industry and AT&T, although he explores the rise of the telegraph in great detail. For a broader look at the institutionalization of American society and the rise of modern bureaucratic organizations see Robert H. Wiebe, *The Search for Order, 1877–1920* (New York, 1967), and David A. Noble's less benign *America by Design: Science, Technology, and the Rise of Corporate Capitalism* (New York, 1977). The above works, along with John A. Garraty, *The New Commonwealth, 1877–1890* (New York, 1968), provide a good overview of the social, economic, political, and institutional contexts for the era.

Some more specialized histories that informed this study are: Milton Friedman and Anna J. Schwarz, *A Monetary History of the United States* (Princeton, 1963); Lawrence M. Friedman, *A History of American Law* (New York, 1973); Oscar Handlin and Mary Handlin, *Commonwealth: A Study of the Role of Government in the American Economy: Massachusetts, 1744–1861* (Cambridge, Mass., 1954); U.S Bureau of the Census, *Historical Statistics of the United States: Colonial Times to 1970* (Washington, D.C., 1975); Gary John Previts and Barbara Dubis Merino, *A History of Accounting in America: An Historical Interpretation of the Cultural Significance of Accounting* (New York, 1979); and S. Paul Garner, *Evolution of Cost Accounting to 1925* (Montgomery, Ala., 1954).

Just as telephony ab initio was an outgrowth of telegraphy and a branch of the emerging electrical industry, it was also, at bottom, a manufacturing business. Victor S. Clark, *History of Manufactures in the United States, 1860–1914* (Washington, D.C., 1928), details the history of manufacturing firms by industrial category. Glenn Porter and Harold C. Livesay, *Merchants and Manufacturers: Studies in the Changing Structure of Nineteenth-Century Markets* (Baltimore, 1971), deals with the rise of vertically integrated firms in terms of changes in the way they brought their products to market. Like Chandler, Porter and Livesay give great weight to the creation of a national market and to advances in technology as an agrarian, merchant-dominated economic order gave way to a more industrialized economy. It was in the midst of this change that firms like the American Bell Company and Western Electric forged their mutual destinies. U.S. Bureau of the Census, 1880, *Report on the Manufactures of the United States at the Tenth Census* (Washington, D.C., 1883), provides a statistical snapshot of this context.

For the electrical industry, general studies are Howard Passer, *The Electrical Manufacturers, 1875–1900: A Study in Competition, Entrepreneurship, Technical Change, and Economic Growth* (Cambridge, Mass., 1953); Arthur A. Bright, Jr., *The Electric Lamp Industry: Technological Change and Economic Development from 1800 to 1947* (New York, 1949); and the excellent "systems" analysis of the electric power industry, Thomas P. Hughes, *Networks of Power: Electrification in Western Society, 1880–1930* (Baltimore, 1983). The standard work on the organization of Western Union is Robert L.

Thompson, *Wiring a Continent: The History of the Telegraph Industry in the United States, 1832–1866* (Princeton, 1947). Thompson's history carries the story through the consolidation of the industry under Western Union's aegis. The monopoly period of Western Union's history has no good historical analyses. See also a specialized piece by Richard B. DuBoff, "Business Demand and the Development of the Telegraph in the United States, 1844–1860," *Business History Review* 54 (Winter 1980), 459–79. Alvin F. Harlow, *Old Wires and New Waves* (New York, 1936), offers some important detail in his chronicle of telecommunications from the telegraph through the radio. A muddled but useful contemporary account is James D. Reid, *The Telegraph in America* (New York, 1879).

Finally, no student of the history of industrial organization can ignore the microeconomic literature on the subject. It is a large and growing field—as much in need of the empirical and detailed analyses of institutional processes that historians can provide as historians themselves are in need of more structured frameworks for the organization of data on particular cases. Two books that have served as useful points of reference for this history are F. M. Scherer, *Industrial Market Structure and Economic Performance* (Chicago, 1980); and Oliver E. Williamson, *Markets and Hierarchies: Analysis and Antitrust Implications* (New York, 1975).

Notes ❧

1 Overview and Context

1. There is a brief summary of the acquisition in the Federal Communication Commission's *Investigation of the Telephone Industry in the United States* (Washington, D.C., 1939), the final report of the commission's lengthy investigation of the industry that began in 1934. Supporting the report are a number of detailed exhibits containing material relating to the early AT&T corporate history, Western Electric, patents, and technical development. Complete sets are available at the FCC, at the Baker Library of the Harvard Business School, and at AT&T. A chapter-length narrative of the history of the Bell Company's pre-1879 manufacturing arrangements is contained in R. J. Tosiello, "The Birth and Early Years of the Bell Telephone System, 1876–1880" (Ph.D. diss., Boston University, 1971). Neither the FCC materials nor Tosiello, however, provides any critical analysis of the early manufacturing strategies as they evolved. Other important works on the early history of the industry, such as N. R. Danielian's *AT&T: The Story of Industrial Conquest* (New York, 1939) and, more recently, Gerald W. Brock, *The Telecommunications Industry: The Dynamics of Market Structure* (Cambridge, Mass., 1981), attempt to characterize the acquisition of Western Electric in terms of their theoretical understanding of economic behavior but provide no empirical evidence of the decision-making behavior of the Bell entrepreneurs. Oddly, John Brooks's centennial history of the Bell System, *Telephone: The First Hundred Years* (New York, 1975, 1976), neglects the company's early manufacturing history.

2. In economic language, the "vertical structure" of a business enterprise is the organization of functions along the axis from the extraction or procurement of raw materials to the delivery of finished products to

the final consumer. In the following discussion of the development of the Bell telephone business, the term "vertical integration" refers to the acquisition and coordination of any two functions along the vertical axis. From the perspective of the founders of the business, who mediated between the production of telephone hardware and the distribution of telephones and telephone service, "backward integration" involved the control of sources of production and supply, and "forward integration" involved the control of distribution and service.

3. Alfred D. Chandler, Jr., *The Visible Hand: The Managerial Revolution in American Business* (Cambridge, Mass., 1977), esp. chap. 9. Chandler's synthesis of the history of American business management and corporate organization has become the point of departure for discussions of the history of management in both the United States and abroad (see Alfred D. Chandler, Jr., and Hermann Daems, eds., *Managerial Hierarchies: Comparative Perspectives on the Rise of the Modern Industrial Enterprise* [Cambridge, Mass., 1980]; and Chandler's earlier, seminal work, *Strategy and Structure: Chapters in the History of the American Industrial Enterprise* [Cambridge, Mass., 1962]).

4. Alfred D. Chandler, Jr., "The United States: Seedbed of Managerial Capitalism," in Chandler and Daems, eds., *Managerial Hierarchies*, 23.

5. In addition to Chandler, *Visible Hand*, 292–97, the especially useful studies include Patrick G. Porter, "Origins of the American Tobacco Company," *Business History Review* 43 (Spring 1969), 59–76; and Reese W. Jenkins, *Images and Enterprise: Technology and the American Photographic Industry, 1839–1925* (Baltimore, 1975), esp. chaps. 4–6.

6. Chandler, *Visible Hand*, 299–302. See also Mary Yeager Kujovich, "The Refrigerator Car and the Growth of the American Dressed Beef Industry," *Business History Review* 44 (Winter 1970), 460–82; and Thomas C. Cochran, *The Pabst Brewing Company* (New York, 1948).

7. Chandler, *Visible Hand*, 302–12; Andrew B. Jack, "The Channels of Distribution for the Innovation; the Sewing Machine Industry in America," *Explorations in Entrepreneurial History* 9 (February 1957), 113–41; Glenn Porter and Harold C. Livesay, *Merchants and Manufacturers: Studies in the Changing Structure of Nineteenth-Century Markets* (Baltimore, 1971), 184–92; and Harold C. Passer, *The Electrical Manufacturers, 1875–1900: A Study in Competition, Entrepreneurship, Technical Change, and Economic Growth* (Cambridge, Mass., 1953).

8. Chandler, in "The United States," 25–26, notes that mass producers, particularly processors of agricultural products, integrated backward

to sources of supply, but most of the cases that he cites illustrate a pattern of manufacturers or processors moving forward into distribution.

9. Ibid., 20–21.

10. On Standard Oil see Ralph W. Hidy and Muriel E. Hidy, *Pioneering in Big Business* (New York, 1955).

11. Ibid., 40–46. The trust was a legal expedient for surmounting the problems of creating a holding company (which required an act of a state legislature). Through the trust, conceived by S. C. T. Dodd, Standard Oil's attorney, a number of firms could exchange their stock for trust certificates issued by a board of trustees; the trustees were then empowered to act as financial and operating managers of the constituent firms. By the 1890s, general incorporation laws in many states obviated the need for trusts (see Chandler, *Visible Hand,* 319). The American Bell Company of 1880, which was empowered to hold stock in other companies, had its origins in a special act of the Massachusetts legislature.

12. Here I am following Chandler's interpretation, which gives great weight to the impact of the technological innovation of the pipeline (see *Visible Hand,* 321–25).

13. For background on the telegraph industry see Robert L. Thompson, *Wiring a Continent: The History of the Telegraph Industry in the United States, 1832–1866* (Princeton, 1947).

14. Throughout, I use *Bell Company* or simply *Bell* to denote the several corporate predecessors of the American Telephone and Telegraph Company.

15. The Articles of Association of the Bell Telephone Company, 29 June 1878, are quoted in [William Chauncey Langdon,] *The Early Corporate Development of the Telephone* (New York, 1935), 15.

16. Alexander Graham Bell to the Capitalists of the Electric Telephone Company (in London), 25 March 1878, quoted in John E. Kingsbury, *The Telephone and Telegraph Exchanges: Their Invention and Development* (London, 1915), 92.

17. Throughout, I shall use *telephone equipment* to refer generically to any device used in conjunction with the telephone or telephone exchange. *Telephone* or *telephone instruments* will refer to only the terminal receiver and transmitter, which were leased by the Bell Company to its operating agencies. *Telephone apparatus* will refer to equipment used in conjunction with, but not including, the telephone receiver and transmitter. These usages are in keeping with distinctions made by the early Bell owners and managers before 1882. After 1882, *apparatus* seems to have been the generic term. The 1882

manufacturing contract uses *appliances* to refer to what had previously been called *apparatus* (see appendix E).

18. U.S. Bureau of the Census, *Historical Statistics of the United States: Colonial Times to 1970* (Washington, D.C., 1975), 12, 224, 913.

19. For a good, brief discussion of the general business and economic contexts of the 1870s see John A. Garraty, *The New Commonwealth, 1877–1890* (New York, 1968), chap. 3. The best authority on the long-term deflation of prices in the 1870s is Milton Friedman and Anna J. Schwartz, *A Monetary History of the United States* (Princeton, 1963), chap. 2. On the rise of business associates and cartels see Chandler, *Visible Hand*, 316–17: "By the 1880s these federations had become part of the normal way of doing business in most American industries."

20. On the early development of the electrical industry in the United States see Victor S. Clark, *History of Manufactures in the United States, 1860–1914* (Washington, D.C., 1928), chap. 32; Arthur A. Bright, Jr., *The Electrical Lamp Industry: Technological Change and Economic Development from 1800 to 1947* (New York, 1949), esp. 32–33; Passer, *The Electrical Manufacturers*, introduction and chap. 1; Thomas P. Hughes, *Networks of Power: Electrification in Western Society, 1880–1930* (Baltimore, 1983), chap. 2; and U.S. Bureau of the Census, 1880, *Report on the Manufacturers of the United States at the Tenth Census* (Washington, D.C., 1883). The quoted phrase is from Clark, *History of Manufactures*, 381. Both Danelian, *AT&T*, chap. 5; and David A. Noble, *America by Design: Science, Technology, and the Rise of Corporate Capitalism* (New York, 1977), chaps. 6 and 7, deal with the relationship between corporate organization and patent rights in the electrical industries generally and telephony specifically. Neither, however, ties the exploitation of patents to particular forms of organization; they dwell instead on what they view as the problems of corporate co-optation of scientific research for monopoly control of the marketplace.

21. For example, in Massachusetts in 1878 only 520 of more than 10,000 manufacturing concerns were corporations. Two-thirds of the goods produced were turned out by nonincorporated manufacturers (see Garraty, *The New Commonwealth*, 97). The corporate device was not generally used for raising capital. In most cases, large interests expanded through combination rather than by floating stock issues.

22. For a fine impression of this prebureaucratic, preprofessional age see Robert H. Wiebe, *The Search for Order, 1877–1920* (New York, 1967), chap. 1.

23. The immediate effects of sociopolitical regulatory forces, so dominant in the environment of the modern Bell System, were negligible in the very early years of telephony. The early Bell System was shaped in a relatively permissive political atmosphere.

24. Cf. Thomas P. Hughes, *Networks of Power*, chap. 1. In his treatment of electrical power systems in Germany, England, and the United States, Hughes stresses the possibility of multiple outcomes in the business development of technology. He makes a persuasive case for the importance of context and styles of management—in addition to the forces of economics and technology—in the development of business-technology systems.

2 Manufacturing Arrangements of the Bell Company to 1879

1. See the deposition of Thomas Watson in *Proofs by and about Alexander G. Bell, 1875–89*, comp. J. J. Storrow, 2 vols. (n.p., n.d.), 2:253, in the American Telephone and Telegraph Company Archives, 550 Madison Avenue, New York, N.Y. (hereafter cited as AT&T Archives).

2. Thomas A. Watson, *The Birth and Babyhood of the Telephone* (New York, 1913), 25.

3. See the compilation "Early Telephone Statistics, 1876–1879," prepared by S. L. Andrew and C. W. Foss (1929), AT&T Archives, box 1006.

4. Sidney H. Aronson, "Bell's Electrical Toy: What's the Use? The Sociology of Early Telephone Use," in *The Social Impact of the Telephone*, ed. Ithiel de Sola Pool (Cambridge, Mass., 1977), 26. Aronson quotes a circular issued by a Bell licensee in Titusville, Pennsylvania, on 27 October 1877. The quote from Watson is from Thomas A. Watson to R. W. Hawkinson, 23 April 1878, copy in R. B. Hill's Papers, vol. 7, AT&T Archives, box 2047. See also Kingsbury, *Telephone and Telephone Exchanges*, chap. 8.

5. Gardiner Hubbard, "The Telephone," May 1877, AT&T Archives, box 1097.

6. Kingsbury, *Telephone and Telephone Exchanges*, 66, 207–8; Aronson, "Bell's Electrical Toy," 26–28.

7. There is some evidence that the patentees looked forward to the day when they could have equity in operations in New England, but it is clear that the initial burden "of investing a large amount in wires and other property" was to be borne by the agents (see Gardiner G. Hubbard to E. T. Holmes, 30 July 1877, AT&T Archives, box 1057).

8. There is no biography of Hubbard. The best readily available information on his life career is in the *Dictionary of American Biography;* and in Robert V. Bruce, *Bell: Alexander Graham Bell and the Conquest of Solitude* (Boston, 1973), 82–87, 90–91, and passim.

9. Bruce, *Bell.* Hubbard did not anticipate the telephone and indeed thought of Bell's obsession with the possibility of transmitting articulate speech electrically as a frivolous distraction from more important work on a multiple telegraph that might transmit several messages over a single wire simultaneously (see David Hounshell, "Elisha Gray and the Telephone: The Disadvantages of Being an Expert," *Technology and Culture* 16 [April 1975], 151).

10. The patents were Alexander Graham Bell's 7 March 1876 grant on the basic principle of telephony (U.S. Patent No. 174,465) and his 30 January 1877 grant on the telephone hardware (U.S. Patent No. 186,787).

11. Bruce, *Bell,* 258–59.

12. These patents, if deemed to "control" the use of telephone technology, would legally secure the holders of the patents from commercial competition until 1894. Others could use the patents for commercial purposes only if licensed to do so by the patentees. Those trying to make or sell a patented device without a license were liable for "infringement" and subject to serious financial penalties.

13. See Sanders to Hubbard, 28 January, 27 February, and 19 March 1878, AT&T Archives, box 1193. "It is pretty hard to show what the actual assets of the company are," Sanders wrote on 19 March.

14. Hubbard to Watson, 23 May 1877, ibid.; "The Telephone."

15. Thomas A. Watson, *Exploring Life* (New York, 1926), 160–61. This was also Hubbard's opinion. See also Sanders to Hubbard, 18 June 1878, AT&T Archives, box 1193.

16. Watson, *Exploring Life,* chap. 4. The size of Williams's shop was impressive considering that in 1880 manufacturers of electrical apparatus averaged 10.5 employees and producers of telephone and telegraph apparatus averaged 22.3 employees. By 1880 Williams had more than 200 employees. The largest manufactories were iron and steel plants, which averaged 140 employees per firm (see U.S. Bureau of the Census, 1880, *Report on the Manufacturers of the United States at the Tenth Census,* 5 ff.; Memo, Draft Telephone Massachusetts Manufacture, Etc., 18 February 1880, AT&T Archives, box 1055).

17. Watson, *Exploring Life,* 31, 52, 86, 127.

18. Tosiello, "Birth and Early Years," 149–55; George Bartlett Prescott, *The Electric Telephone,* 2d ed., rev. (New York, 1890), 439–43;

Frederick Leland Rhodes, *Beginnings of Telephony* (New York, 1929), 178–79.

19. Kingsbury, *Telephone and Telephone Exchanges,* 77–82, traces the ideal of interconnecting telegraph lines "for public use through a central office" to Francois Dumont, of Paris, in 1851 (under a British patent). Telegraphic exchange systems were established in both Great Britain and the United States many years before the telephone was invented. Bell licensees experimented with exchange systems throughout 1877.

20. Prior to that time, switchboards were fashioned on site by the licensees who operated them. Williams's first switchboard was based on a design created by Thomas B. Doolittle, who operated an agency in Bridgeport, Connecticut (see Doolittle's unpublished manuscript "My Work in the Telephone Field," typescript, AT&T Archives, box 2026).

21. Agreement between the Bell Telephone Company and Charles Williams, Jr., 1 August 1878, copy in AT&T Archives, uncatalogued. For the complete text of the contract see appendix E.

22. The Bell Company apparently leased and furnished call bells itself (after buying them from Williams) before this new contract took effect (see the Bell Telephone Company "Circular Letter" to Agents, 2 September 1878, AT&T Archives, box 1123). Thomas Sanders attributed much of the financial distress of the Bell Company to "the necessity of furnishing Magneto Bells . . . , which Mr. H. refused to sell but insisted upon leasing . . ." (see Sanders to [A. G.] Bell, 29 September 1878, ibid., box 1003).

23. See Hubbard to Watson, 17, 21, and 28 November 1877, ibid., box 1193. See also Watson to Hubbard, 19 November 1877, General Manager's Letter Books, vols. 11–177 (1877–84), 11:436–47. These Letter Books, which contain copies of outgoing correspondence relating to day-to-day operations, are on microfilm in the AT&T Archives. Hereinafter they are cited as GMLB.

24. Trustee's Report to the Board of Managers, [August] 1877, AT&T Archives, box 1001.

25. Hubbard to Watson, 11 April 1878, ibid., box 1193, reported the following figures for the production of telephones and call bells. It is obvious that twice (in December and March) Williams achieved production of more than an average of 50 telephones per day (based on a six-day work week and including one Christmas holiday for December). Otherwise he fell short, averaging only about 41 in January 1878. Williams's production from August 1877 through March 1878 was as follows:

	Telephones	Call Bells
Before August 1877	800	20
August 1877	660	140
September 1877	760	186
October 1877	840	315
November 1877	1,000	420
December 1877	1,420	676
January 1878	1,060	380
February 1878	920	390
March 1878	1,346	411

26. Tosiello, "Birth and Early Years," 140 ff., covers this well. See also Sanders to Hubbard, 3 and 28 January and 19 March 1878, AT&T Archives, box 1193. See also Williams to Hubbard, March 1878.

27. See above, sec. II, and below, sec. IV.

28. Bell Telephone Company, "Instructions to Agents, No. 3," 1 February 1878, AT&T Archives, box 1001. This is the earliest instance of encouragement to licensees to push the exchange business that I know of.

29. Ibid.

30. Bruce, Bell, 260–61; Brooks, Telephone, 61–62.

31. Ibid. See also Brock, The Telecommunications Industry, 93; and James D. Reid, The Telegraph in America (New York, 1879), 629–33.

32. See Sanders to Hubbard, 5 December 1877, and Charles Cheever to Hubbard, 3 December 1877, AT&T Archives, box 1006; "Diary of First Trip Made by Thomas A. Watson among the Agents of the Bell Telephone Co.," 1 March 1878, ibid., box 1069; Sanders to Hubbard, 22 February 1878, ibid., box 1193. Sanders perceived immediately the problem of attracting new investors in the face of Western Union's superior resources. His suggestion that the Bell interests consider selling out appears in Sanders to Hubbard, 30 January 1878, GMLB.

33. Correspondence between Sanders and Hubbard during this period is found in the AT&T Archives, box 1193. This paragraph draws in particular on five letters from Sanders to Hubbard: 19 and 29 January, 27 February, and 1 and 5 March 1878.

34. Hubbard to Cheever, 5 February 1878, ibid., box 1006. According to Bruce, Bell, 262, negotiations (at least the first round) were broken off on 21 February 1878. They resumed later in the year.

35. When Orton died in April, Sanders reflected: "A powerful man in his position with his animus toward Mr. Hubbard I am convinced has worked us a great injury. He never meant that GGH should make one dollar if he could help it" (Sanders to Cheever, 23 April 1878,

GMLB, 13:420). Orton's dislike of Hubbard may have orginated in Hubbard's suggestions in 1869 that the telegraph industry be placed under the direction of the Postmaster General's Office (see *The Postal Telegraph: The Only Means by Which the Telegraph Can Be Made the Ordinary Method of Communication. An Address Given by G. G. Hubbard before the Board of Trade and Commercial Exchange*, Philadelphia, 25 November 1869, copy in AT&T Archives; see also Bruce, *Bell*, 127, 229). A bill based on Hubbard's proposal had come close to passage in the U.S. Congress in 1874.

36. For a detailed account of the Bell Company's problems with Western Union competition, see Tosiello, "Birth and Early Years," chaps. 9 and 10, which I have followed closely here.

37. On the early history of Western Electric see Charles G. DuBois, "A Half Century of Western Electric Achievement," *Western Electric News* 8 (November 1919), 1–6. A very useful but less accessible summary is found in FCC, Hearings before the Telephone Division, March 1936–June 1937, "Report on Western Electric Company Incorporated Corporate Structure, Manufacturing Facilities and Cost Accounting System," exhibit 1952, 5 February 1937, pp. 3 ff.

38. DuBois, "A Half Century of Achievement."

39. Ibid.

40. Ibid. There had been only seventeen employees at Gray and Barton in 1870. See also the unsigned memorandum on Western Electric employees, AT&T Archives, box 1046.

41. Ibid.

42. See Watson, *Birth and Babyhood*, 4: "When a piece of machinery built by the Western Electric came into our shop . . . we . . . always used to admire the superlative excellence of the workmanship."

43. Anson Stager to Hubbard, 15 July 1877, AT&T Archives, box 1151. Copies and some originals of correspondence relating to Western Electric–Bell Company relations during 1877–78 are stored in the AT&T Bell Laboratories Archives, AT&T Bell Laboratories, Murray Hill, New Jersey.

44. Enos Barton to Stager, 20 October 1877, AT&T Archives, box 1151. "We have gone on the same principle in regard to telephones that we follow in other matters, that in the business which we get we should try to give our customers what they will consider an equivalent for their money; so that we have to admit that thus far this business in the city of Chicago has not been a success, and this is not for lack of soliciting but because (of) the interference on the lines. . . ." The problem of "interference," which seemed especially nettlesome in

Chicago, is discussed in M. D. Fagan, ed., *History of Engineering and Science in the Bell System* (n.p., 1975), 203–4. Barton actually suggested the use of a metallic circuit, which anticipated the solution to the problem of long-distance transmission from Boston to Providence in 1881.

45. Barton to Hubbard, 22 December 1877; and Stager to Hubbard, 28 December 1877, both AT&T Archives, box 1151.

46. Stager to Hubbard, 14 January 1878, ibid.

47. Hubbard to Sanders, 28 April 1878, ibid., box 1193. See also "Diary of First Trip by Thomas A. Watson," ibid., box 1069; and Watson to Hubbard, 2 March 1878, in the Western Electric collection at the AT&T Bell Laboratories Archives.

48. Tosiello, "Birth and Early Years," 280 ff.

49. Barton to Hubbard, 1 June 1878, AT&T Archives, box 1151. Barton reiterated the proposal a few days later (Barton to Hubbard, 6 June 1878, ibid.), saying specifically that the American District Telephone Company would pay the Bell Company four dollars each for the latter's telephones in exchange for the granting of a license under Bell patents, "whereby they could use other telephones than yours, upon payment to you at a royalty of, say, three dollars on each telephone. . . ."

50. Hubbard to Barton, 10 June 1878, ibid. This followed by six days a letter in which Hubbard had offered to receive royalties for the use of Bell patents (Hubbard to Barton 4 June 1878, ibid., box 1195).

51. H. H. Eldred to Hubbard, 5 June 1878, ibid., box 1152; Cheever to Hubbard, 3 December 1877, ibid., box 1006.

52. Tosiello, "Birth and Early Years," 406, 408–9, 415–16.

53. Bell Telephone Company to C. H. Haskins, 10 August 1878, AT&T Archives, box 1131. The district exchange system involved transfers of messages among the telephone, telegraph, and courier. District exchanges did not necessarily involve physically interconnected switching systems for the direct transfer of speech among the various telephone subscribers to the exchange.

54. Watson to Sanders, 28 May 1878, ibid., box 1193.

55. Tosiello, "Birth and Early Years," chap. 8, covers the organization of the New England Company in great detail. For a brief treatment see J. Warren Stehman, *The Financial History of the American Telephone and Telegraph Company* (Boston, 1925), 8–12.

56. Stehman, *Financial History*, 8, cites the first reason as the basis for the organization of the New England Company. Tosiello, "Birth and Early Years," carries a running account of Hubbard's desire in par-

ticular to keep Bell patents closely held. On the patentees' interest in purchasing agency properties in New England see again Hubbard to Holmes, 30 July 1877, AT&T Archives, box 1057: "We shall probably . . . organize a company to [take financial] control [of] the whole business for New England. . . ." On the impending sale of the Boston agency to the Bell interests see George Bradley to Hubbard, 5 January 1878, ibid., box 1031.

57. Tosiello, "Birth and Early Years," 219–21. It is worth noting that the Bell agency in New York was not an exchange operation, which is no doubt another reason why it could not compete well with Western Union's Gold and Stock exchange (see the reminiscence of William A. Childs, "Law Telephone System, 1877–1888"; and Childs to Thomas J. Perkins, 12 December 1915, AT&T Archives, box 1055).

58. Tosiello, "Birth and Early Years," 308.

59. Albert Bigelow Paine, *Theodore N. Vail: A Biography* (New York, 1921), 107–12; Watson, *Exploring Life,* 142.

60. Paine, *Theodore Vail,* 4–5, 14–32, 58–92, 197–112. See also the scattered biographical materials on Vail in the AT&T Archives, box 1080.

61. See B. C. Forbes, *Men Who Are Making America* (New York, 1916), 377. In an interview with Forbes, Vail noted that he "was to get $5,000 (per annum) when he could collect it—which was seldom!" Watson wondered whether Hubbard could meet Vail's high salary demands (*Exploring Life,* 142).

62. Tosiello, "Birth and Early Years," 221.

63. [Langdon,] *The Early Corporate Development,* 15, copy in AT&T Archives, box 1116.

64. Tosiello, "Birth and Early Years," 286 ff.; Stehman, *Financial History,* 8–12.

65. Rhodes, *Beginnings of Telephony,* 76–79; Vail to Watson, 8 September 1878, AT&T Archives, uncatalogued; Frederic William Wile, *Emile Berliner: Maker of the Microphone* (Indianapolis, 1926), 107 ff.; Tosiello, "Birth and Early Years," 344–56. The Blake transmitter, although clear and articulate, later proved too low in volume and power for long-distance transmission (see Fred DeLand, "Notes on the Development of Telephone Service," *Popular Science Monthly* 70 [May 1907], 403–6.

66. When the Bell Company refused Edison's offer in April 1878, it was largely because of its faith in Watson's ability to produce a transmitter that would overcome the problems of Alexander Graham Bell's design. Watson himself, after initially opposing the idea, proposed

that the Edison transmitter be purchased. Hubbard came to the same conclusion too late (see Hubbard to Sanders, 25 March and 20 April 1878, AT&T Archives, box 1193).

67. See Watson to Hubbard, 26 March 1878, ibid.; and Madden to Vail, 30 November 1878, and Vail to Madden, 22 January 1879, both in ibid., box 1197. There were probably no more than two or three traveling agents by 1879, but if their salaries (not including expenses) were equivalent to Madden's, they were handsome by contemporary standards.

68. Tosiello, "Birth and Early Years," 761–67. Bell's keen interest in Chicago was based on Hubbard's firm belief that it would become the best market in the country for telephony (see Hubbard to the Executive Committee, 3 and 10 August 1878, ibid., box 1152). The Boston correspondence is spotty, but see Holmes (the Bell agent for Boston) to Bradley, 29 March 1878, ibid., box 1031.

69. See "Bell, Licensees, October 1877–January 1905," ibid., box 1011; and *Record Book: American Bell Telephone Company Agreements*, vol. 1, original in AT&T Archives.

70. I am indebted to Robert W. Garnet, of AT&T, for sharing his findings on early Bell Company strategic and structural relations with its operating licensees. See also Vail to E. W. Gleason, 25 October 1878, GMLB, in which the general manager explains what had by then become standard policy for contracting licensees.

71. Tosiello, "Birth and Early Years," 363–66, discusses the tangled financial arrangements between Bell and its Chicago exchange. I have not been able to ascertain whether it was necessary at this stage for Bell to receive an explicit grant from the state allowing it to hold stock in other companies. Such power was explicitly granted in the 1880 charter for the American Bell Telephone and by that time was apparently needed as a matter of politics and law. Compare the charters of the Bell Telephone Company (30 July 1878, AT&T Archives, box 1003), and the American Bell Telephone Company (19 May 1880, ibid., box 100).

72. Sanders to Vail, 15 November 1878, and Bradley to Vail, 13 November 1878, both in ibid., box 1194.

73. See above, sec. II.

74. See above, ibid. and n. 26.

75. Williams to Hubbard, 17 December 18[78], AT&T Archives, box 1204 (incorrectly dated 1887 by Williams). Earlier in the year, when Hubbard was contemplating licensing another manufacturer, Williams wrote, explaining: "I have taken additional rooms, put in

additional machinery, made changes and increased my running expenses expressly for your work. Furthermore I can do the manufacturing to better advantage and cheaper than any establishment can" (See Williams to Hubbard, 7 March 1878, ibid.).

76. See letters from George Bradley to various correspondents from 12 February through 21 May 1878 in George Bradley's Letter Books, 2 vols. (1878), 1:32–33, 45, 70, 77, 92, and 308, AT&T Archives. See also Sanders to Hubbard, 21 January and 15 March 1878, AT&T Archives, box 1193 (exhibit 12). Two telephones had been required at each station partly because of weaknesses in transmission. Technical improvements and soaring demands brought pressure from the agents to drop the requirement.

77. Sanders to Hubbard, 18 June 1878, and Sanders to A. O. Morgan, 15 June 1878, ibid., box 1193.

78. This became especially true after the introduction of the Blake transmitter. Early on, Bell adjusted prices, altered various licensing discounts, and provided financial assistance to critical franchises. By early 1878, however, it preferred to compete more on the strength of its patent claims and on the reputation of its instruments. Watson, in a letter to Vail dated 19 February 1879 (ibid., box 1205), reported that subscribers using both Western Union and Bell services were finding that they received from Bell quicker answers to calls, better telephones, and better-made equipment more promptly repaired.

79. Watson to Hubbard, 12 April 1878, ibid., 14 April 1878, ibid., box 1193. Watson did say, however, that there might be a need for establishing relations with other manufacturers in order to lighten Williams's load.

80. Hubbard to the Executive Committee of the Bell Company, 18 August 1878, ibid., box 1152. Vail to Williams, 18 September 1878, GMLB. Durant to Watson, 27 December 1878; Durant to the Bell Telephone Company, 11 February 1879; and Durant to Vail, 6 March 1879, all in AT&T Archives, box 1164.

81. See G. C. Maynard to Vail, 3 October 1878, ibid., box 1178; T. E. Cornish to Vail, 1 February 1879, ibid., box 1176; E. J. Hall to Vail, 10 January 1879, ibid., box 1164. Maynard was agent for the District of Columbia, and Cornish was agent for Philadelphia. Complaints about call bells were not new, however, having surfaced as early as July 1877 (see the correspondence from licensees, ibid., box 1170; and M. K. Applebaugh to Watson, 15 October 1877, ibid., box 1168). It is worth noting that complaints about telephones seem to have related more to problems generic to the technology than to

quality of manufacturing. Statistically, from June 1877 through December 1878, 8.6 percent of telephones shipped to agents were returned. What portion of that percentage represents equipment failure as opposed to simple replacement of older by newer and better instruments is unknown. See "Instruments Shipped to and Returned by Agents and Exchanges in the United States," appended to "Recapitulation: United States Instruments," 20 February 1880, ibid., box 1004.

82. Williams to Vail, 19 December 1878; and Williams to Watson, 7 February 1879, ibid., box 1204.

83. Ibid. See also Fagan, ed., *History of Engineering and Science*, 69–74, 88–90.

3 Manufacturing under the License Contracts, 1879–1881

1. Alexander Graham Bell to Mabel Bell, 24 January 1879, Bell Papers, National Geographic Society, quoted in Tosiello, "Birth and Early Years," 379.

2. See Tosiello, "Birth and Early Years," 372 ff.

3. Ibid., 403 ff.

4. Ibid.

5. See above, chap. 2.

6. Tosiello, "Birth and Early Years," 377 ff.; [Langdon,] *The Early Corporate Development*, 17–21; Arthur S. Pier, *Forbes: Telephone Pioneer* (New York, 1953), 117–22.

7. Stehman, *Financial History*, 18–19.

8. Tosiello, "Birth and Early Years," 368–72. Hubbard favored the New York location, which he felt would give the Bell Company more national stature and better access to new investors. The result was to separate operations from executive ownership (in a way that Vail found troublesome) and from production.

9. Ibid., 380 ff.; [Langdon,] *The Early Corporate Development*, 19.

10. Pier, *Forbes*, 11–17, 34–61, 92.

11. See Vail to John Ponton, 9 April 1879, AT&T Archives, box 1123; Vail to Hawkins, 25 April 1879, GMLB; and letter of authorization for J. M. Brown, 30 June 1879, GMLB.

12. See *A Description of the Telephone Exchange System . . .*; *Instructions for Establishing and Operating an Exchange* (Boston, 1880), bound in "Lockwood's Notations," a collection of early telephone pamphlets on various operational and technical matters in the AT&T Archives.

The pamphlet is a virtual how-to-do-it manual on establishing and operating an exchange.

13. See Vail to Cornish, 20 June 1879, AT&T Archives, box 1054, and Vail to G. S. Glen, 29 May 1879, GMLB, on the company's intentions to provide long-distance service.

14. On the Bell Company's competitive tactics see Tosiello, "Birth and Early Years," 410 ff. See also Vail to Hall, 26 April 1879, GMLB.

15. Vail to Watson, 17 and 18 February 1879, AT&T Archives, box 1205; Vail to Williams, 21 February 1879, GMLB 20:101. See also Watson to Vail, 25 February 1879, AT&T Archives, box 194: "Durant [St. Louis] complains of the delay in receiving goods from Williams. I have got him to order a sample magneto from Post and Co. and he will probably use their bell hereafter."

16. Watson to Vail, 28 February 1879, AT&T Archives, box 1205. At this point Vail was thinking in terms of one or two manufacturers in addition to Charles Williams, Jr.—one in New York and one in the West. Other company officials were beginning to agree.

17. Bradley to Forbes, 12 April 1879, ibid., box 1195.

18. See Doolittle, "My Work in the Telephone Field."

19. See the R. G. Dun and Company Collection, Baker Library, Harvard University, Cambridge, Massachusetts, for contemporary credit ratings on the Bell Company manufacturing licensees of 1879. It appears that R. G. Dun and Company's credit reports on the manufacturing firms licensed by Bell were generally favorable, with the possible exception of that on the Electric Merchandising Company. Relevant volumes are: Maryland, vol. 15 (Davis and Watts); Illinois, vol. 44 (Electric Merchandising Company); Indiana, vol. 3 (Indianapolis Telephone Company); Ohio, vol. 8 (Post and Company); and Massachusetts, vol. 84 (Charles Williams, Jr.).

20. Partrick and Carter was one of many firms doing business with the Bell Company for nonpatented apparatus. In 1878 it had requested samples of a double-acting ground-switching telephone holder and an annunciator. It may have produced one or more of these items. From 1877, however, there is correspondence indicating a chronic shortage of funds (see Partrick and Carter to Hubbard, 3 October 1877, 10, 19, and 26 April and 5 October 1878; and Partrick and Carter to Sanders, 15 August 1879—all AT&T Archives, box 1202).

21. Hubbard to Vail, 1 March 1879, ibid., box 1194.

22. Hubbard to Vail, 14 April 1879, ibid., box 1151; West and Bond to the National Bell Telephone Co., 29 April 1879, ibid., box 1195.

23. Davis and Watts to Vail, 29 January 1879, ibid., box 1158.

24. E. T. Gilliland to Watson, 14 April and 8 June 1879, ibid., box 1153. Gilliland had left Post and Company with some hard feelings for unspecified reasons. His indictment of the Post and Company president, E. V. Cherry, ran deeper. "Cherry's idea," he wrote on 14 April, "is that after the exchanges in the large cities get started there will be nothing more to do and they are making all the money they can out of it while it lasts." As for the "unsafe" facility, there was apparently a fatal industrial accident in June.

25. Post and Company to Vail, 23 January, 15, 19, and 23 May 1879, ibid., box 1203.

26. See Madden to Vail, 31 May 1879, ibid., box 1197. Gilliland's business was principally financed by W. S. Morrison, president of the First National Bank of Indianapolis, and William O. Rockwood, who had extensive holdings in local insurance, manufacturing, and railroad companies. Morrison and Rockwood were president and treasurer, respectively, of the corporation (see the correspondence relating to the Indianapolis Telephone Company, ibid., box 1153).

27. Gilliland to Watson and Gilliland to Madden, both 14 April 1879, ibid.

28. On the Gilliland switchboard see Prescott, *The Electric Telephone*, 288; Kingsbury, *Telephone and Telephone Exchanges*, 164–65; and *The Third Meeting of the National Telephone Exchange Association* (New Haven, 1881), 41. In 1880 Vail wrote to Forbes saying that Gilliland "has done more towards developing the apparatus used in conjunction with Exchanges than any man . . ." (see Vail to Forbes, 14 December 1880, GMLB).

29. See Gilliland to Watson, 1 May 1879, AT&T Archives, box 1153; Madden to Vail, 31 May 1879, ibid., box 1197. The then contemporary Western Electric technician H. B. Thayer emphasized Gilliland's "cheap methods of manufacture" in "The Development of Telephone Manufacturing," 5–6, ibid., box 1045.

30. Madden to Vail, 31 May 1879, ibid., box 1197; Watson to Gilliland, 11 June 1879, GMLB; C. B. Rockwood to Vail, 17 July 1879, AT&T Archives, box 1153; and Rockwood to National Bell Telephone Co., 14 August 1879, ibid.

31. The correspondence throughout the spring indicates the Bell Company's preference for keeping the telephone manufacturing exclusively with Charles Williams, Jr. Vail to Post and Co., 22 April 1879, GMLB, distinguishes between "telephone apparatus," which Post might be allowed to produce, and transmitters, which were to be shipped to agents by the Bell Company. For Vail's final decision see Vail to E. H. Parker, 11 July 1879, GMLB.

32. Copies of the manufacturing license contracts and supplemental agreements can be found in the AT&T Archives, uncatalogued. For the complete text of the Electric Merchandising Contract see appendix E.
33. Cf. above, chap. 2, sec. II.
34. The equity of treatment of the manufacturers becomes apparent from a scanning of operational correspondence in GMLB.
35. Watson to Gilliland, 11 June 1879, GMLB.
36. Vail to Post and Co., 22 April 1879, ibid.
37. See above, sec. II and n. 16. The quote is from Post and Company to Vail, 23 May 1879, AT&T Archives, box 1203.
38. See Vail to Manufacturers, 15 December 1880, GMLB; and Vail to Gilliland, 9 May 1881, ibid. Contrast this relatively liberal approach toward the auxiliary apparatus with the determination to control the technological development of the telephone in one shop under a "very rigid inspection" (see Vail to Post and Co., 5 May 1879, ibid.).
39. Vail to Post and Co., 12 August 1879, ibid.
40. On nineteenth-century cost accounting see S. Paul Garner, *Evolution of Cost Accounting to 1925* (Montgomery, Ala., 1954); H. Thomas Johnson, "Early Cost Accounting of Internal Management Control: Lyman Mills in the 1850s," *Business History Review* 46 (Winter 1972), 466–74; and American Telephone and Telegraph Company, *Depreciation History and Concepts in the Bell System* (New York, 1957), chap. 1.
41. Williams to Hubbard, 13 March 1878, AT&T Archives, box 1204.
42. Bradley to Hubbard, 15 April 1878, George Bradley's Letter Books, 1:236, AT&T Archives; Sanders to Hubbard and Hubbard to Sanders, both 15 March 1878, AT&T Archives, box 1193. See also, Tosiello, "Birth and Early Years," 141–42, 212–13.
43. A. O. Morgan to Sanders, [June 1878], AT&T Archives, box 1193; Madden to Vail, 28 November 1878, ibid., box 1197; Williams to Watson, 28 and 31 December 1878, ibid., box 1204.
44. Watson to Vail, 21 March 1879, ibid., box 1205.
45. The cost of making a telephone in 1877 was $2.86. It dropped to as low as $2.29 from January to April 1878 before rising to its year-end cost of $2.70. On the cost of telephones and magneto bells in early 1878 see Hubbard to Watson, 11 April 1878, ibid., box 1193.
46. See Madden to Vail, 31 May 1879, ibid., box 1197.
47. *Report of the Joint Committee of the Senate and Assembly of the State of New York Appointed to Investigate the Telephone and Telegraph Companies,* vol. 1 (Albany, 1910), 418–19 (hereinafter cited as *New York Investigation, 1910*).

48. See, for example, Vail to Davis and Watts, 23 January 1880, GMLB.
49. See three sets of letters from Vail to manufacturers, 1 November and 21 December 1880 and 7 January 1881, ibid. The nine exchanges were located in Boston; New York City; Chicago; Galveston, Texas; Elmira, New York; San Francisco; Houston; Philadelphia; and Canada (precise whereabouts unspecified). See also Vail to Pacific Bell Telephone Co., Telephone Dispatch Co. et al., 7 January 1881, GMLB; Vail to Gilliland, 31 January 1881, ibid.; and Williams to Agents and Exchanges, circular letter, 17 December 1880, AT&T Archives, box 1071.
50. The acceptance of this discriminatory pricing might otherwise be attributed to Vail's insistence that the matter of discounts to privileged exchanges be kept "strictly confidential" (see again Vail to Manufacturers, 7 January 1881, GMLB; and Vail to George Phillips, 7 January 1881, AT&T Archives, box 1131). But it is highly unlikely that this would have remained a secret.
51. The only direct complaints that I have seen about the level of pricing came from an agent in Cincinnati, who objected strenuously to an inventor's royalty charged by the Bell Company after it had settled its patent suit with Western Union in November 1879, and from Post and Company in early 1880. The Cincinnati agent protested that the cost of the royalty was being passed on from the manufacturers to the agencies. Post and Company complained that the agencies were making handsome profits, while wholesale prices were kept too low in relation to production costs (see Vail to W. H. Eckert, 1 December 1879, GMLB; Vail to Post and Company, 10 December 1879, GMLB; and Post and Company to Vail, 30 January 1880, AT&T Archives, box 1117).
52. See an explanation of this change in policy in Vail to Post and Co., 19 January 1880, GMLB. The royalties to be paid to the inventors were fixed in a contract struck between the Bell Company and various inventors, effective 1 June 1879. The manufacturers were to absorb this cost as of 12 November 1879, two days after the execution of the Bell–Western Union agreement. The Bell Company obviously did not want to risk any disturbance of its manufacturers prior to the settlement (see also Vail to W. O. Rockwood, 25 July 1879, copy in R. B. Hill's Personal Papers, vol. 7, AT&T Archives, box 2047). Post and Company had suggested that the cost of royalties be divided between the Bell Company and the manufacturers, but the Bell Company preferred to raise the product price (see Madden to Post and Co., 22 November 1879, GMLB).

53. Memorandum [unsigned] prepared for Annual Report of the American Bell Telephone Company, March 1880, AT&T Archives, box 1004.

54. Memorandum, Vail to Forbes, March 1881, ibid., box 1007.

55. For various perspectives on the settlement see Brooks, *Telephone*, 71–72; Tosiello, "Birth and Early Years," chap. 14; Julius Grodinsky, *Jay Gould* (Philadelphia, 1957), 206; FCC, Investigation, 123–24; and Michael F. Wolff, "The Marriage That Almost Was," *IEEE Spectrum*, February 1976, 41–51.

56. I have relied on a copy of the affidavit of George Gifford, dated 19 September 1882, in *American Bell Telephone Company v. John J. Ghegan et al.*, AT&T Archives, uncatalogued.

57. This line of argument has been developed most fully in FCC, *Investigation*, exhibit 2096F, 13–31.

58. Vail to D. H. Ogden, 26 May 1879, GMLB.

59. The initial proposal offered by Western Union suggested the creation of a new company in which the Bell interests and Western Union were each to hold stock, the allocation of which was to be settled by arbitration (see Forbes to Hubbard, 25 May 1879, President's Letter Books, vol. IG, AT&T Archives).

60. FCC, *Investigation*, exhibit 2096F, 13.

61. Tosiello offers a detailed description of the Ormes-Louderback agreement in "Birth and Early Years," 470–80.

62. Vail to Forbes, 30 July 1879, AT&T Archives, box 1034.

63. See the printed version of the agreement entitled "Contract" 10 November 1879, ibid., box 1006, hereinafter cited as Contract. Copies of the signed agreement are located in the Western Electric collection at the AT&T Bell Laboratories Archives.

64. Stehman, *Financial History*, 19.

65. Western Union and its subsidiary, the Gold and Stock Company, owned fifty-five telephone exchanges at the time of the settlement. Western Union held an interest in eighteen more, five of which were Bell licensees. Twelve additional independent exchanges used Gold and Stock telephones. All of these would either become Bell licensees or have their plant and equipment absorbed by Bell licensees (see Schedules B, C, and D of the agreement in Contract).

66. Norvin Green to Vail, 3 September 1879, ibid., box 1006.

67. See J. Bunnel to Vail, six letters dated 5 June 1879 through 8 January 1880; and H. W. Pope to Vail, 7 January 1880—all AT&T Archives, box 1202. The Madden letters are Madden to J. G. Elwood, 8 January 1880; and Madden to Durant, 29 November 1880, both GMLB.

68. Vail, Report on Operations of the Telephone Business, 19 March 1880, typescript, AT&T Archives, box 1080. General Manager's Report, 10 March 1880, ibid., box 1007, gives the projection. The actual number of telephones produced between 1 March 1880 and 20 February 1881 was 66,762. The number of telephones actually placed in the hands of agents in the United States increased by about 42,000 (see appendix A).
69. See appendix A.
70. Cf. "Instruments Shipped to and Returned by Agents and Exchanges in the United States," which notes 34,964 net shipments of telephones for 1879, or 672 per week.
71. See Contract, Thomas A. Watson and the National Bell Telephone Company, 1 May 1879, ibid., box 1070; and Emile Berliner to Paul Magne and Henry Pelletier, 17 December 1879, GMLB.
72. On Gilliland see above and C. B. Rockwood to the National Bell Telephone Company, 6 September 1879, AT&T Archives, box 1153. See also Williams to Vail, 28 January 1880, ibid., box 1024; E. V. Cherry to Watson, 8 November 1879, ibid., box 1203; and Post and Co. to Watson, 22 December 1879, ibid.
73. Vail to Post and Co., 22 April 1879, GMLB; Vail to Davis and Watts, 21 May 1879, ibid.
74. On the Post and Company–Indianapolis Telephone Company conflict see Post and Co. to Vail, 9 August and 4 September 1879, and Post and Co. to Watson, 3, 22, and 29 December 1879 and 24 January 1880, all AT&T Archives, box 1203; C. B. Rockwood to Vail, 14 and 21 August and 6 September 1879, and Rockwood to Watson, 4 December 1879, ibid., box 1153. Rockwood pleaded "self protection" in face of the "sharp" competition from Post and Co., which, he argued, the Bell Company ought not to permit. See also Electric Merchandising Co. to Madden, 28 November 1879; Electric Merchandising Co. to Watson, 15 December 1879; and Post and Co. to Vail, 30 January 1880, all ibid., box 1151.
75. Madden to Post and Co., 2 March 1880; Madden to Indianapolis Telephone Co., 16 April 1880; and Vail to Post and Co., 15 April 1880 and 13 July 1882, all GMLB.
76. Vail to Post and Co., 29 June 1882; and two sets of letters from Madden to Manufacturers, 12 November 1880, all ibid.
77. C. N. Fay to Watson, 14 and 19 January 1880, AT&T Archives, box 1151.
78. Post and Co. to Vail, 30 January 1880, ibid., box 1117; W. H. Harrington to Bradley, 2 August 1879, ibid., box 1151; and Electric Manufacturing Co. to Watson, 15 December 1879. ibid.

79. This parallels somewhat Vail's concern that when agencies impinged on one another's territory, the effects were "demoralizing" (see Vail to Post and Co., 22 March 1880, GMLB).

80. See esp. George Bliss to Vail, 29 November 1880, and Tompkins to Vail, undated, copies in AT&T Archives, uncatalogued. At the same time that Bliss was accusing the Western Electric Manufacturing Company of selling unlicensed magneto bells to the Lockwood Telephone Company, an informant told Vail that the Electric Merchandising Company was itself offering magnetos to unauthorized parties. Examples from GMLB are Vail to Williams, 8 March 1880; Vail to Manufacturers, 25 August 1880; Madden to Manufacturers, 18 September 1880; Vail to Manufacturers, 27 October 1880; Vail to Post and Co., 2 November 1880; Madden to Post and Co., 30 December 1881; and Madden to Manufacturers, 21–26 May and 29 June 1881.

81. In GMLB see Vail to Williams, 3 January, 30 July, and 4 August 1880; Vail to Davis and Watts, 15 January 1881; Vail to Manufacturers, 12 June 1882; and Vail to Post and Co., 29 August 1882.

82. See, for example, Vail to Electric Merchandising Co., 14 April 1880; Vail to Manufacturers, 21 April 1880; and Vail to Davis and Watts, 8 May 1880, all in GMLB.

83. Charles E. Scribner to Hall, 10 January 1906, AT&T Archives, uncatalogued. Scribner began his career at Western Electric in the late 1870s. His negative comments on the quality of early Bell telephone apparatus may reflect a kind of company loyalty, since there is little direct evidence to support such a strong indictment of the 1879 manufacturing licensees. In fact, at least one, Post and Company, was considered to have an excellent line of call bells, whose reputaton surpassed those of Western Electric's even after the acquisition.

84. Deposition of Theodore N. Vail, *Western Union v. American Bell Telephone Co.*, exhibit C, "Evidence for the Defendant," 2 vols. (Boston, 1909), 2:1556; C. E. Scribner, in *Western Electric News*, August 1927, 15–17.

4 Strategies for Vertical Control and the Acquisition of Western Electric, 1880–1882

1. See above, chap. 3, sec. VI and n. 65.

2. See article 7 of Contract.

3. Forbes to H. L. Higginson, 24 April 1880, AT&T Archives, box

1055; General Manager's Report, 1880, ibid., box 1007. In fact, the output of telephones by Charles Williams, Jr., between 1 March 1880 and 20 February 1881 was 66,762. The number of telephones in the hands of Bell licensees increased from about 91,000 to more than 132,000 during the same period. About 29,000 telephones in the hands of domestic licensees were Gold and Stock products (see appendix A).

4. Frank B. Jewett, "The Telephone Switchboard—Fifty Years of History," *Bell Telephone Quarterly* 7 (July 1928), 156 ff. Operators "could handle the calls for one hundred to two hundred subscribers" by making "interconnections between their boards by reaching across in front of each other." This, of course, had its limits in larger exchanges which required more boards, more operators, and awkward trunking systems.

5. The Bell Company provided careful instructions on this point in *Description of the Telephone Exchange System*, 24–26. In the 1880s more problems of induction interference arose with the emergence of the light and power industries.

6. Looking back years later, Vail wrote (somewhat defensively) about the vulnerability of Bell's patent position: "While the settlement with the Western Union Telegraph Company in 1879 removed from the field the most formidable and powerful competitor, it must not be concluded that the American Bell Telephone Company had the field to itself. . . . Patents and inventions were necessary for defense but there was not protection against imitators. There was a continued running fight in the courts and in the field. The fact that Bell won every case in the courts availed it nothing except that it was credited with a monopoly that did not exist" (see the *1909 Annual Report of the Directors of the American Telephone & Telegraph Company to the Stockholders* [Boston, 1910], 21–22).

7. See the schedules appended to Contract.

8. On the legal history of the Eaton, Dolbear, and Drawbaugh cases, which set in train a series of trials and appeals resulting in the Supreme Court's narrow 4–3 ruling upholding the validity of the Bell patents in 1888, see Brooks, *Telephone*, 76–81.

9. On switchboards see Prescott, *The Electric Telephone*, chap. 8; Kingsbury, *Telephone and Telephone Exchanges*; Rhodes, *Beginnings of Telephony*, 150 ff.; and Fagan, ed., *History of Engineering and Science*, 482–95. The standard Gilliland and Williams boards, the most popular of the Bell manufacturers' models in the early 1880s (see *The Fourth Meeting of the National Telephone Exchange Association* [Chi-

cago, 1883], 38–39), had capacities of 50–75 circuits. The 1880 catalogue of the Charles Williams, Jr., shop, uncatalogued, advertised a "standard" board for large offices with a capacity of 75 circuits. The Law switchboard allowed for the connection of a single board and was easily trunked to other boards in an exchange office. The Western Electric "multiple switchboard," which became the basis for most large-scale switchboard development in the latter part of the century, was the preferred alternative.

10. Vail's 1908 remarks are quoted in Danielian, *AT&T*, 96. For a more contemporary usage see Vail to J. J. Storrow, 23 July 1881, GMLB. See also Vail to Forbes, 27 March 1880, AT&T Archives, box 1195, in which Vail asked for authorization to purchase patents routinely up to a maximum of $1,000 and to pay royalties up to $5,000 on instruments actually put into use.

11. Letter from Forbes, 17 September 1880, copy in AT&T Archives, uncatalogued (emphasis added; the name of the addressee is illegible). See also Forbes to Higginson, 24 April 1880, ibid., box 1055. "The control of the field which we are rapidly securing . . . will soon be as valuable to us as our patent rights." These letters provide particularly succinct statements of a strategy that had been unfolding for many months.

12. These points are inferred largely from the logic of the contemporary situation as it emerges from the primary sources. For a somewhat comparable view see Brock, *The Telecommunications Industry*, 100.

13. Text of talk by T. N. Vail, dated 1 November 1915, AT&T Archives, box 1031.

14. On this point see letter from Hubbard (addressee unknown), dated 1878 (possibly 1879), ibid., box 1115. Hubbard surmised that the technical "difficulties in the use of the Telephone on long lines arising from earth currents, atmospheric electricity, retardation of the currents & from other disturbing influences: might someday be obviated," but it was clear "that ultimately the chief use of the Telephone on long lines will be for the transmission of *telegraphic* messages." The telephone was conceived, therefore, as simply a better means for transmitting written intelligence, that is, as a "substitute for the Morse instrument," involving less operator skill and providing more rapid transmission of messages. This view of the potential market for long-distance telephony persisted through the 1880s.

15. A few months before the settlement, Vail had advised at least two licensees that the Bell directors were contemplating "connecting cities and towns" by wire (see Vail to Cornish, 20 June 1879, ibid.,

box 1054; and Vail to Glen, 29 May 1879, GMLB). It is possible that Vail was referring to telegraph connections or to a combination of local telephone and long-distance telegraph. The letter to Glen mentions telephones, but the context is not clear. Construction of a telephone line between Boston and Lowell, a distance of thirty miles, was begun in late June. Other correspondence of Bell officials points to an interconnected exchange business that would compete with Western Union's. See, for example, [Sanders] to the Stockholders of the Bell Telephone Company, [January 1879,] AT&T Archives, box 1003: "As the business developes [sic] and the various district systems established in all parts of the country become connected by wires and the transmission of messages for hire is actively engaged . . . , the sympathy and support of the public and local capital will no doubt furnish these connecting lines on such terms as will give us the control of a system which will be a most powerful opposition to the W.U."

16. Memo, "Draft Telephone Massachusetts Manufacture Etc., February 18, 1880" (in Forbes's hand), ibid., box 1055.

17. See above, chap. 2, sec. VII; and chap. 3, sec. I.

18. Vail to W. K. Rice, 1 September 1879, GMLB. See also Vail to Maynard, 13 September 1879, ibid.: "Our idea, now is to provide additional capital, so that we can take part interest in any exchange . . . , leaving the management of such exchanges to local parties. . . ."

19. Draft Report to the Stockholders of the National Bell Telephone Company, 1879, AT&T Archives, box 1195. This report may have been prepared for a special meeting of the stockholders held on 24 October 1879. It is more likely that it was written for the end-of-the-year report given in December. See also Vail to Maynard, 13 September 1879, GMLB.

20. See the unsigned memoranda "Important Work *Immediately* before the National Bell Telephone Company," 15 December 1879; and "Draft Report to the Stockholders," both in AT&T Archives, box 1195.

21. See Stehman, *Financial History*, 22–23.

22. On the legal background see Lawrence M. Friedman, *A History of American Law* (New York, 1973), 446 ff.; and Oscar Handlin and Mary Handlin, *Commonwealth: A Study of the Role of Government in the American Economy: Massachusetts, 1744–1861* (Cambridge, Mass., 1954). The central problem for the Bell Company was the contemporary application of the doctrine of ultra vires in corporate law, that is, the proposition that the corporation was of limited authority

and that any action taken by the corporation not specifically authorized in its corporate charter was prohibited. Massachusetts was
especially stringent about granting rights to corporations, especially
those engaged in mining and manufacturing. National Bell's capitalization of $850,000 was probably in the upper range for manufacturing
firms in Massachusetts in the 1870s.

23. [J. J. Storrow,] Memorandum, January 1880, AT&T Archives, box
1326. The tone of the memorandum suggests that it was part of an
apologia for the charter petition.

24. Ibid. "[The] good of the public . . . requires us to grant exclusive
territorial rights coupled with an obligation to be diligent in extending the use of telephones by furnishing them to all."

25. Commonwealth of Massachusetts, An Act to Incorporate the American Bell Telephone Company, AT&T Archives, box 1007.

26. The quoted language is from testimony given by Forbes on 9 March
1886 ("Testimony for Remonstrants at Hearings before State Legislature Committee on Mercantile Affairs, re: Various Bills to Regulate
Telephone Charges . . . ," copy of excerpts of Forbes's testimony in
AT&T Archives, uncatalogued). In fact, American Bell received
conflicting legal advice on the matter but proceeded on the liberal
course (see the opinions of Elias Merwin and C. W. Bradley, April
1880, AT&T Archives, box 1007).

27. Stehman, *Financial History,* 20–22; [Langdon,] *The Early Corporate
Development,* 22–24.

28. See AT&T Co. and Predecessor Companies—Directors and Officers,
1877–1956, AT&T Archives, box 1010; General Manager to the
President of the National Bell Telephone Company, 1 November
1879, ibid., box 1005. Salaries accounted for nearly half of the
operating expenses of the head office.

29. See the "Chronological List of Officers and Employees of the American Bell Telephone Co. and Its Predecessors. Before May 1902,"
ibid., box 1008. The engineer was Joseph P. Davis, who received his
civil engineering degree from Rensselaer Polytechnic Institute and
went on to become Bell's chief engineer in the 1890s. He was hired for
his expertise in underground construction, and his first task was to
help design conduits for wire for Bell exchange operations in New
York City ("Joseph P. Davis—Pioneer in Underground Telephone
Construction," *Bell Laboratories Record* 26 [November 1948],
457 ff.).

30. Initially the agents were to "establish a correct, uniform and prompt
system of reports" (see Vail to C. J. French, 16 May 1879, GMLB).

The growing responsibilities of the traveling agents is tracked in great detail by James P. Baughman in written testimony for *U.S. v. AT&T*, delivered in January 1982. My copy of Baughman's testimony was provided by the legal department of AT&T, 550 Madison Avenue, New York, N.Y. See pp. 33–37.

31. On the Electrical and Patent Department see Vail to Thomas D. Lockwood, 5 February 1881, GMLB; and the memorandum from Thomas Watson on the activities of the Electrical and Patent Department (1880–81), reprinted in Federal Communications Commission, Special Investigation Docket No. 1, exhibit 1951A, app. D.

32. See Stehman, *Financial History*, 23–24.

33. On the company's attempts to develop long-distance lines between major urban centers through independent contractors see Baughman's testimony (cited above, n. 30), 38 ff.

34. See above, chap. 3, sec. VII.

35. Williams to National Bell Telephone Company, 2 March 1880, AT&T Archives, box 1204.

36. Ibid.

37. Madden to Williams, 3 March 1880, GMLB.

38. Memorandum, 14 July 1880, AT&T Archives, box 1282.

39. See above, chap. 2, sec. IV.

40. [Sanders] to the Stockholders of the Bell Telephone Company, [January 1879,] AT&T Archives, box 1003; Forbes to R. S. Fay, President's Letter Books, 9 May 1879. It is hard to account for the alleged vast difference in cost, unless Sanders and Forbes were dealing with apples and oranges. Unlike Bell telephones, Gold and Stock instruments were packaged and shipped with call bells and other apparatus in a single set. Still, Bell officials were convinced of a significant difference in their and Western Union's manufacturing costs.

41. See Clark, *History of Manufactures*, chap. 32 and p. 381; and U.S. Bureau of the Census, 1880, *Report on the Manufacturers of the United States at the Tenth Census*, 1327–40.

42. This is plausible when we note that as of 1877 there were only 12,250 "instruments in use" under control of the Western Union Company, including "Morse sounders," "Morse recorders," printers, and various multiplex instruments (see Reid, *The Telegraph in America*, 575). It is hard to imagine that the rate of growth of telegraph instruments, most of which required skilled operators, would have reached a rate of manufacture equal to Charles Williams, Jr.'s annual output of 57,000 telephones in 1879–80.

43. Milo Kellogg to Barton, 25 July 1879, Enos M. Barton papers, in the Western Electric collection at the Bell Laboratories Archives; and D. H. Louderback to Vail, 12 July 1879, and Vail to Louderback, 15 October 1879, both in AT&T Archives, box 1202.

44. See the text of a lecture by L. S. O'Rourk, "The History of the Western Electric Company, 1869–1924," 17 October 1924, ibid., box 2069.

45. See the memoir by Charles Scribner in *Western Electric News* 1 (February 1913), 4; *Western Union v. American Bell Telephone Co.*, "Evidence for the Defendant," 1:784. Western Electric continued to work repairing equipment and augmenting those telephone exchanges using Gold and Stock telephones. With its switchboard patents, it could hope to secure business in large exchange offices. The firm was also free to produce and sell a variety of electrical apparatus used in the telephone plant that fell outside the controlling influence of the Bell patents. But aside from a one-shot order from Bell for 800 magnetos (see Vail to Western Electric Manufacturing Co., 21 April 1880, GMLB), Western Electric enjoyed none of the fruits of this highly lucrative new business in call bells. Moreover, its residual supply, maintenance, and repair work for the old Western Union exchanges, all of which were being absorbed by Bell licensees, were rapidly diminishing.

46. The best brief account of Stager's hopes for Western Electric in the 1870s is the L. S. O'Rourk lecture of 1924 (see above, n. 44). Western Electric had also sold its rights to another potentially valuable product—the typewriter—to Remington for $10,000.

47. For evidence of the ties between Western Electric and the midwestern telephone exchanges after the Western Union–Bell agreement of 1879 see Madden to the Western Electric Manufacturing Co., 16 and 17 January 1880; and Madden to Kellogg, 5 February 1880, all GMLB. See also correspondence relating to the organization of the United Telephone Company of Missouri and the Western Telephone Company, 1880–81, AT&T Archives, box 1216.

48. See Barton to Kellogg, 28 May 1880, Barton papers.

49. The total per annum sales from 1873 to 1882 are reported in the Western Electric Manufacturing Company Secretary's Report, Chicago, 15 May 1882, AT&T Archives, box 1216. It is also interesting to note the perspective of one early Western Electric salesman, H. F. Albright, at the turn of the decade: "The decade from 1869–1879 was the small shop period marked by the usual groping to find profitable lines of manufacture. Our affiliation with the telephone industry

furnished the needed main line" (see Albright's memoir in *Western Electric News* 8 [November 1919], 25–26).

50. See Vail to Forbes, 1 March 1882, transcript, AT&T Archives, uncatalogued. See also Western Electric Company, Employees, 1883; Barton to Vail, 20 May 1882; and Secretary's Report, 15 May 1882—all in ibid., box 1216.

51. Rhodes, *Beginnings of Telephony*, 154 ff.; testimony of E. M. Barton, *Western Union v. American Bell Telephone Co.*, "Evidence for the Defendant," 1:696–98.

52. Rhodes, *Beginnings of Telephony*, 104, 126. *The Third Meeting of the National Telephone Exchange Association*, 18–36, reveals that by 1880 "public authorities of several of the larger cities" were pressing for legislation "to relieve the streets of unsightly telegraph supports." Cables, as everyone in the telephone business knew, were crucial to the extension of urban communications in the 1880s.

53. See the copy of the contract between J. H. Irwin and the Western Electric Manufacturing Co., 20 January 1880, AT&T Archives, box 1046.

54. See Vail to J. H. Irwin, 22 January, 12 February, and 24 May 1880, all GMLB. See also Forbes to Storrow, 19 May 1880, President's Letter Books. Forbes urged a meeting with Irwin.

55. Barton to Kellogg, 5 February 1880, Barton Papers. Barton's unhappiness with Western Union's involvement in the Irwin-Voelker matter was expressed thus: "It seems to me pretty rough for them [Western Union] to give away their hold on the Induction Coil in telephone transmitters . . . and all that, and then make us poor chaps assume $10,000 a year, and the expenses of a patent litigation on the Irwin-Voelker lay, in order to force the Bell Telephone Co. to come to terms in the four cities. . . ."

56. Barton to Kellogg, 28 May 1880, Barton Papers.

57. Vail to Western Electric Manufacturing Co., 7 June 1880, GMLB.

58. Gilliland's business relationships extended not only to Western Electric but even to T. T. Eckert, Gould's associate, who would soon take over management of Western Union. Gilliland was also investing his money in telephone exchanges (see the materials relating to the Gilliland Company in AT&T Archives, box 1153).

59. The *Indianapolis News*, 3 January 1880, reported that Gilliland was a stockholder in the combined Western Union–Bell agencies in Indianapolis "and will probably have a prominent place in its management. . . . The Western Electric exchange of Chicago have [sic] a large interest in the new venture. . . ."

60. See the sales pamphlet *Western Electric Company* (Indianapolis, 1881), in the Western Electric collection at the Bell Laboratories Archives. Gilliland had delivered 4,950 magneto bells in the first quarter of 1880, which, given the 90,000–95,000 telephones in the hands of Bell licensees (see appendix A), was a very substantial rate of output. The Gilliland Company also boasted 735 of its "standard" switchboards sold and in use.
61. See above, chap. 3, sec. III.
62. See above, chap. 2, secs. IV and V.
63. On Gould's takeover of Western Union see Grodinsky, *Jay Gould*, 146–57, 202–6, 259–88, for the most complete treatment. See also Matthew Josephson, *The Robber Barons*, 2d ed. (New York, 1962), 205–8.
64. The Gould takeover of Western Union as a stimulus to Western Electric's alliance with the Bell Company has been cited in Horace Coon, *American T&T: The Story of a Great Monopoly* (New York, 1939), 119; and Joseph Goulden, *Monopoly* (New York, 1968), 37–39, 81. See also Thayer, "The Development of Telephone Manufacturing," AT&T Archives, box 1045; and *Lindheimer v. Illinois Bell*, 292 U.S. 151 (1934), testimony of Thomas Lockwood, transcript of record, 1:453. Thayer and Lockwood, who had been working for Western Electric and the Bell Company, respectively, in 1881, both recalled the fears of Western Electric stockholders and employees.
65. Unsigned memorandum, 20 April 1881, AT&T Archives, box 1216.
66. Western Electric's nominal capitalization was $300,000.
67. See appendix D.
68. See the summary of correspondence between Bell and Western Electric for the late spring and summer of 1881 in AT&T Archives, box 1046.
69. Ibid. The quote is from Williams to Vail, 23 June 1881, copy, AT&T Archives, box 1046.
70. *Report of the Directors of the American Bell Telephone Co. to the Stockholders*, 29 March 1882, 6, AT&T Archives.
71. Details of the stock transactions can be traced through the small body of papers held in the Western Electric collection at the Bell Laboratories Archives. For information on the brief period during which Western Electric was held in trust, I am grateful to Alan Gardner and Laurence Steadman, who called to my attention their research into the legal history of American Bell as a holding company. The 21 May 1883 revision of its charter by Chapter 200 of the Acts of 1883 was repealed in 1886, but the repeal did not affect acquisitions of stock

taken in the interim. Subsequent reforms of Massachusetts corporate law then eliminated the problem for American Bell altogether.

72. Of the Bell Company's remaining manufacturing licensees, one was the Detroit Electrical Works, which by an old agreement struck by Gardiner Hubbard in 1877 was allowed to make call bells for the Michigan agency (see appendix D). The Detroit Electrical Works was already controlled by Western Electric. The Electrical Merchandising Company was failing as a business before 1882. Davis and Watts managed to delay the termination of their contract for a while, but it was finally severed in August 1883. As for Post and Company, its license was terminated, but a special arrangement was made to continue the production of its finely regarded call bells through a subsidiary partially owned by Bell. Materials relating to the fate of the licensees can be found in GMLB and in AT&T Archives, boxes 1151, 1211, 1215, and 1216.

5 The Aftermath: Consequences of the Acquisition, 1882–1915

1. The quote is from Edgar S. Bloom, a Western Electric president during the 1930s (see *Lindheimer v. Illinois Bell Tel. Co.*, 292 U.S. 151 [1934], transcript of record, 1:409).
2. See 1882 manufacturing contract between the American Bell Telephone Company and the Western Electric Company, 6 February 1882, secs. 1 and 2. (The original contract can be found in the office of the Secretary, AT&T, 550 Madison Avenue, New York City.) For the complete text of the contract see appendix E.
3. Ibid., secs. 3 and 4.
4. Ibid., sec. 4(j).
5. Ibid., secs. 4(h) and 5.
6. Ibid., section 4(e) (emphasis added).
7. Deposition of Enos M. Barton, *Western Union v. American Bell Telephone Co.*, "Evidence for the Defendant," 2:706.
8. Deposition of Henry B. Thayer, ibid., 2:666–67. "We have also had to sacrifice other lines of manufacture. In the periods in which the telephone companies were less busy we have tried to fill up our shops with other lines of work, and then, as the telephone companies became very busy again, we have had to turn work away. Very likely that has had to do with the fact that our business has become so largely a telephone business."
9. In 1879, Thomas Watson had written to E. T. Gilliland saying that

the latter's Bell business would "only be limited by the capacity of your factory." It was, said Watson, Theodore Vail's hope that Gilliland "guarantee to rush things and turn out instruments rapidly." But neither the Bell Company's manufacturing contracts of 1879 nor its agreement with Charles Williams, Jr., provided formal assurances for capacity on demand (see Watson to Gilliland, 11 June 1879, GMLB).

10. Barton to F. P. Fish, 16 March 1907, AT&T Archives, box 1380.

11. See Barton to Vail, 21 January 1882, and Vail to Barton, 25 January 1882, both in ibid., box 1216.

12. On Western Electric's entry into the field of power generation see Barton to Vail, 3 October 1884; Kellogg to Vail, 1 November 1884; W. S. Smoot to American Bell Telephone Co., 14(?) July and 3 August 1885; Smoot to Forbes, 13 August 1885; and Smoot to Vail, 15 August 1885, all in ibid., box 1239. Smoot, president of Western Electric following Anson Stager's death in 1884, insisted that the company was not in the business of "running plants," a business that he found "very distasteful," but that it was necessary to place plants in operation to create the market for manufacturing electric lights.

13. See the scattered correspondence on the subject of incandescent lamps in ibid., box 1216, through 1891; Barton to Fish, 16 March 1907, ibid., box 1380; and advertising circulars labeled "Western Electric Company and Representative Types of Western Electric Arc Lamps," ibid.

14. C. G. DuBois to Vail, 11 December 1907; Thayer to Vail, 14 and 17 December 1907; and Barton to Vail, 5 October and 14 December 1907; all in ibid., box 1380.

15. On the sale see Dubois, "A Half Century of Achievement," 5.

16. "I do not think there was anyone in the country, unless it was Mr. Vail, who recognized the real field for the development of the telephone," testified former AT&T president Frederick Fish in a 1913 stockholders suit. Fish reasoned that "if the Bell Company had succeeded in developing the Business . . . in order to meet the real needs of the communities, there would never have been any independent companies at all." A copy of excerpts from Fish's testimony is in the AT&T Archives, uncatalogued.

17. The statistical information is from "Telephone Development in the United States, 1876 to 1957," AT&T, Chief Statistician's Division (June 1958), AT&T Archives, box 1006. See also U.S. Bureau of the Census, *Historical Statistics of the United States: Colonial Times to 1970*, pt. 2, pp. 783–84. I thank Neil Wasserman for calling to my attention his extrapolation of the rate of telephone growth from 1895 to 1907.

18. See H. F. Albright, "Fifty Years' Progress in Manufacturing," *Western*

Electric News 8 (November 1919), 23, 28; and *WE*, September–October 1981, 8–9.

19. Gerard Swope, "The Western Electric Company's Place in the Bell System," Western Electric Company Conference, Hot Springs, Va., 22, 23, 24 April 1913, app. B. copies in AT&T Archives and Bell Laboratories Archives. The two larger companies were Allgemeine Electricitaets Gessellschaft and General Electric, with $86 million and $70 million in sales, respectively. On Western Electric's foreign operations in 1914 see Mira Wilkins, *The Emergence of Multinational Enterprise: American Business Abroad from the Colonial Era to 1914* (Cambridge, Mass., 1970), 51, 200, 213. The first plant was built in Antwerp in 1882, and others followed, largely in response to governments' preference to purchase locally made equipment for nationalized telephone services. At AT&T's behest, Western Electric sold its overseas plants in 1925 in order to concentrate more fully on domestic, Bell System markets.

20. "Telephone Development in the United States."

21. Berliner to Vail, 3 August 1881, photocopy in AT&T Archives, uncatalogued.

22. "Telephone Switchboards," Report of a conference held at the office of the American Telephone and Telegraph Company, 10, 20, 21 December 1887, 160 ff., ibid. See also *Arguments and Testimony . . . as to Telephone Rates and Tolls . . . before the Committee on Mercantile Affairs* (Boston, 1886), 65–66. A 2,700-line multiple switchboard installed in Boston in 1883–84 cost the exchange $40,000 initially plus a patent royalty of $1,500 per annum to Western Electric.

23. "Telephone Switchboards," 250.

24. See the extensive discussions of Western Electric's duties in these regards in Swope, "The Western Electric Company's Place in the Bell System," and in the program of the Western Electric Company's "Manufacturing and Engineering Conference," Chicago, 24–28 May 1915. The latter conference publication is located in the Bell Laboratories Archives. See also H. B. Thayer, *Operating Methods and the Handling of Irregular Calls from the Manufacturer's Point of View* (Boston, 1905), 6, copy in AT&T Archives, box 1139.

25. On the impact of loading see Neil Wasserman, *The Path from Invention to Innovation: The Case of Long-Distance Telephone Transmission at the Turn of the Century* (Baltimore, in press); Thomas Shaw, "The Conquest of Distance by Wire Telephony," in *The Telephone: An Historical Anthology*, ed. George Shiers (New York, 1977), 370 ff.; and James E. Brittain, "The Introduction of the Loading Coil: George

A. Campbell and Michael I. Pupin," *Technology and Culture* 11 (January 1970), 36–57. Useful general discussions of the evolution of research at Western Electric are Leonard S. Reich, "Industrial Research and the Pursuit of Corporate Security: The Early Years of Bell Labs," *Business History Review* 54 (Winter 1980), 504–29; and Lillian Hoddeson, "The Emergence of Basic Research in the Bell Telephone System, 1875–1915," *Technology and Culture* 21 (1981), 512–44.

26. The copper-wire metallic circuit was introduced throughout the network of Bell operating companies gradually over a twenty-year period. It was the *sine qua non* of early long-distance transmission. A metallic circuit was a twisted pair of wires, forming a complete wire circuit, eliminating the ground return used in telegraphy and primitive telephony. It relieved the problem of "earth noise" induction, which interfered with the clear transmission of voice signals. The loading coil reduced the serious attenuation of the electrical impulse as it traveled over a long wire or through an underground cable. These developments can be tracked in Fagan, ed., *History of Engineering and Science*. For some detail on the complex interrelationships among the various parts of the plant and on the systems problems introduced by innovations (such as metallic circuits and loaded lines) see the contemporary correspondence of Bell officials on telephone service improvement and metallic circuits, 1888–92, AT&T Archives, box 1244; and Joseph P. Davis to Fish, 27 January 1903, ibid., box 1360.

27. See the testimony of James Porter Baughman in *U.S. v. AT&T* (1982), "Technological Episode No. 2," A-94. See also Alan Gardner, "Summary of Research on the History of Long Lines, 1894–1927," 21 January 1981, AT&T Archives, 20 ff.

28. Vail to Barton, 3 July 1882, GMLB. On Western Electric's management see R. C. Clowry to Vail, 27 October 1881, copy in AT&T Archives, uncatalogued. Western Electric, Clowry wrote as a prospective stockholder with Bell in the manufacturer, "has prospered in spite of the Management which is known to be inefficient." Vail replied that Clowry's information "corresponds very closely with the opinions we had already formed . . ." (Vail to Clowry, 11 November 1881, copy in AT&T Archives, uncatalogued. See also Vail to Barton, 29 July 1882, GMLB; and Scribner to Hall, 10 January 1906, copy in AT&T Archives, uncatalogued).

29. See the correspondence on the Standard Electrical Works in AT&T Archives, box 1216.

30. Albright, "Fifty Years' Progress," 24. "Some lucky days we got perhaps as high as a dozen or two telephones (of forty-eight) accepted.

The shop superintendent quit in despair, but the shops kept everlastingly at it and at last succeeded in shipping telephones that would stay shipped."

31. Ibid., 25–28.
32. See John J. Carty to Hall, 17 July 1907, AT&T Archives, box 2045. Brief, concurring discussions of the 1907 reorganization appear in Paul R. Lawrence and Davis Dyer, *Renewing American Industry* (New York, 1983), 214; and Fagan, ed., *History of Engineering and Science*, 44.
33. Swope, "The Western Electric Company's Place in the Bell System."
34. See, for example, the standard text, Frederick M. Scherer, *Industrial Market Structure and Economic Performance* (Chicago, 1980), 78.
35. Committee Report to the Board of Directors of the Western Electric Company, attached to Barton to Vail, 2 August 1882, AT&T Archives, box 1216.
36. On this last point especially see the deposition of Henry B. Thayer, *Western Union v. American Bell Telephone Co.*, "Evidence for the Defendant," 2:665.
37. Ibid.
38. Thayer is quoted in Swope, "The Western Electric Company's Place in the Bell System," 10.
39. Alexander Cochrane to James E. Mitchell (vice president, Bell Telephone Company of Philadelphia), 1 March 1901, copy in AT&T Archives, uncatalogued; Thayer to Cochrane, 4 March 1901, copy in ibid.
40. DuBois is quoted in Swope, "The Western Electric Company's Place in the Bell System," 36.
41. Shaw, "The Conquest of Distance," 375 ff.
42. Cf. Robert W. Garnet, *The Telephone Enterprise: The Evolution of the Bell System's Horizontal Structure, 1876–1909* (Baltimore, 1985), on the functional reorganization of AT&T that began in 1907. It was implemented throughout the Bell System gradually over the following decade.
43. Western Electric's move to New York was owing to Illinois's more restrictive laws regarding the conduct of nationwide operations by firms chartered in that state (see *WE*, September–October 1981, 12).
44. On the matter of "transaction cost savings" I am following the insights of Oliver E. Williamson (see his "Emergence of the Visible Hand: Implications for Industrial Organization," in Chandler and Daems, eds., *Managerial Hierarchies*, 182–202, esp. 187, 195; for an extended discussion see idem, *Markets and Hierarchies: Analysis and Antitrust Implications* [New York, 1975]).

Summary and Conclusion

1. See Thomas Hughes's discussion of the phases of development of business-technology systems in his *Networks of Power*, 14–17.
2. Cf. Chandler, *Visible Hand*, 286, 289.
3. Important contemporary cases of technological and organizational integration could be found among the major railroads (particularly the Pennsylvania Railroad) and in the oil business (particularly Standard Oil). In neither business did a single firm combine extensive development of a national market with systematic engineering of technology in as thoroughgoing a manner as did AT&T. By 1915, AT&T was also as sophisticated in its large, functionally organized structure as Du Pont, the case made classic by Chandler, *Strategy and Structure*, chap. 2 and passim.

Appendix C

1. See, for example, Tosiello, "Birth and Early Years," 71–81, which is the most complete summary of the decision.
2. This has been the response of the Bell System in recent years to criticism of its leasing policy.
3. Forbes to George Crocker, 28 September 1880, President's Letter Books, vol. 1G.
4. Chauncey Smith, Jr., to Langdon, 17 November 1923, AT&T Archives, box 1141.
5. Ponton to Bell, 15 June 1898, ibid., box 1123.
6. Forbes to Crocker, 28 September 1880; Hubbard, "The Telephone." Forbes's full statement was as follows: "The principle is that we prefer to provide capital ourselves for the important uses unless others will pay more than we ask for the less important uses. But we find people ready to pay according to use and so we divide the customers into classes and state to them what they and we agree to perform." Separate classes of service for residence and business customers were provided initially, forming the basis of what would later be called "value of service pricing."
7. Madden to W. H. Hamilton, 21 February 1880, GMLB.

Appendix D

1. For correspondence relating to the Michigan Telephone and Telegraph Construction Company see AT&T Archives, box 1159. A

copy of the contract can be found in the AT&T Archives, uncatalogued.

2. Madden to Vail, 6 July 1879; and Jackson to Vail, 3 December 1879, both in ibid., box 1159.

3. See Barton to Hudson, January 1882, ibid., uncatalogued.

Index ❧

George David Smith is president of the Winthrop Group, Inc., a business consulting firm, and he teaches administrative and business history at New York University.

THE JOHNS HOPKINS UNIVERSITY PRESS

The Anatomy of a Business Strategy

This book was composed in Goudy Old Style by Brushwood Graphics Studio, from a design by Sheila Stoneham. It was printed on 60-lb. Warren's Olde Style paper and bound in Holliston Roxite A cloth by the Maple Press Company.